EXTREMES

EXTREMES

LIFE, DEATH AND THE LIMITS
OF THE HUMAN BODY

DR KEVIN FONG

HODDER &
STOUGHTON

First published in Great Britain in 2013 by Hodder & Stoughton
An Hachette UK company

1

Copyright © Kevin Fong 2013

A CIP catalogue record for this title is available from the British Library

Hardback ISBN 978 1 444 73774 5
Trade Paperback ISBN 978 1 444 73775 2
eBook ISBN 978 1 444 73776 9

Typeset by Hewer Text UK Ltd, Edinburgh
Printed and bound by CPI Group (UK) Ltd, Croydon CR0 4YY

Hodder & Stoughton policy is to use papers that are natural, renewable
and recyclable products and made from wood grown in sustainable
forests. The logging and manufacturing processes are expected to
conform to the environmental regulations of the country of origin.

Hodder & Stoughton Ltd
338 Euston Road
London NW1 3BH

www.hodder.co.uk

For Dee, Jack & Noah

CONTENTS

Some of the material in this book draws upon my experience as a doctor. Where necessary the case details of patients I attended, as well as the names of staff involved, have been altered to protect confidentiality.

INTRODUCTION

ASTROPHYSICS: A DEGREE with some stuff about space and a bit of exploration thrown in. I didn't really think about it much harder than that. I was OK at maths, I liked physics and, even though I was constantly being told that it was time to grow up and start making serious decisions about life, something of my childhood fascination with astronauts lingered on. This, and the alphabetical arrangement of the handbook of university courses, helped shape my decision about what to study. It was – like all good exploration – a happy accident.

I didn't really understand what I was in for. The physics department at University College London had to drag me through the course. I used to roll up to the professorial offices immediately after lectures had finished, whining that I hadn't understood a thing and pleading with them to explain it all again. This they often did, patiently sitting me down and talking me through the material for a second – or sometimes even a third – time.

I had a sense of wonder about the subject which was unmatched by my mathematical abilities. I learned to understand the boards full of algebra and equations as though I was stumbling through a foreign language, translating it line by line into something vaguely comprehensible. I could just about grasp the meaning but much of the nuance was lost. It was like a French literature scholar reading *Les Misérables* and understanding, by the end, only that it was a book about a man who went to prison for longer than perhaps he should have.

There was one other thing that started to bother me. In a galaxy of a hundred billion stars, in a universe more than ten billion years old, we appeared to be the only iteration of life – intelligent or otherwise. When it came to biology, looking out past the planets of our solar system, past the Sun, out across the aching chasms of space and on towards the other stars like our own, there was nothing to see but sterility, nothing to hear but silence.

Next to all of those supremely exotic objects, stars composed of neutron soups, points of infinitely curved space and time, cataclysmic explosions – visible to the naked eye across the emptiness – that hurled new elements out across the void, it was life that remained the unanswered question. In among objects that were nearly as old as the universe itself, that spanned the full range of physical extremes from absolute zero to the heat of the Big Bang, there was us: feeble, fragile, fascinating.

~

When I wasn't studying I spent time working for the student newspaper of the University of London. I was supposed to look after the photography but one day the editor let me have a go at writing a story. I had found Professor Maurice H. Wilkins at King's College London, seeing out his final years as an Emeritus Professor in a small office with a view of a brick wall. He, the third awardee of the Nobel Prize for the discovery of the double helix structure of DNA, alongside Francis Crick and James Watson, is often forgotten; and it was about this and his formative experiences that I wanted to ask.

Wilkins was less flamboyant than his collaborators and, in some accounts, enjoyed a difficult working relationship with Rosalind Franklin. He was also a man of socialist values which, as for so many others during the Cold War, was enough to briefly attract the attention of the British intelligence services. I didn't know any of this at the time. Instead I found myself distracted from my shoddy interview questions by the story of his change of track from physics to biology.

Wilkins had been a student of Lawrence Bragg, who himself had been the youngest ever winner of a Nobel Prize – aged only twenty-five – for his work in the field of X-ray crystallography. During the Second World War, Wilkins was selected to work on the Manhattan Project and spent this time at the University of Berkeley in California, investigating uranium isotope separation techniques. Wilkins' eventual contribution to the design and delivery of the first nuclear weapons in Japan was minor, but it appears he was never quite able to forgive himself.

This, he told me, led him to seek an alternative career, away from physics and closer to biology. He would turn his refined techniques of X-ray crystallography, and the magnificent resolving power of those high-energy beams, to the field of biophysics and the dissection of organic molecules. This Wilkins regarded as a move 'from the science of death to the science of life'.

As I packed up my tape recorder at the end of the interview, he asked me what I was studying. When I told him I was in the final year of a degree in astrophysics he smiled and replied: 'Ah yes, I used to be interested in all of that; and then I came back down to Earth.'

Those words struck a chord and so after graduating from astrophysics I went to medical school, deciding to move from the science of dead stuff to the science of staying alive.

∼

I may have come back down to Earth, but the fascination with space never left me. The more I learnt about the human body, about how readily it unravelled in the face of physical challenge, the more amazing it seemed that we continued to roam, pushing at the edges of our narrow envelope of survival, hoping all the time that it would not push back. It seemed nothing short of unbelievable that we would strap human beings to vehicles with the explosive capacity of small nuclear weapons and then shoot them through the layers of our atmosphere to fly around the

Earth or onwards to the Moon.

As a medical student, I sent letter after letter to NASA, hoping for an opportunity to spend my final year elective period with them. I sent faxes from the offices of the student newspaper and eventually bought a dial-up modem and started sending emails. One of these missives finally washed up on a desk somewhere and I was sent an application form for a little-known training rotation at Johnson Space Center in Houston, Texas. I applied and forgot about it. There were only four places available worldwide and they didn't usually take students from overseas.

And then one day an envelope dropped through my letterbox. It had the red stamp of the United States Postal Service and a black and white NASA insignia in the top right-hand corner. I opened it like Charlie opening a bar of chocolate. And inside there was a golden ticket. I was going to Houston.

Tom Wolfe, in his account of the early days of astronautics and the events of Project Apollo, is less than kind about the city of Houston. He talks about the crushing humidity and bone-chilling air conditioning; about the sticky residue of oil refinery gas that seems to hang in the Texas air. These are all fair observations, but what Wolfe doesn't quite capture is the draw that the space centre had, and continues to have, for those with their hearts set upon the adventure of human space flight. To the people who work there, Johnson Space Center is where human space exploration gets done; everyone else is just pretending.

NASA's Lyndon B. Johnson Space Center in Houston is

home to mission control and the astronaut training grounds. We flew parabolic flights, endured explosive decompression, span on chairs and crammed into shuttle simulators. After a month there I knew I had to find a way back. Soon I was volunteering to work unpaid for weeks at a time with the medical operations group, generally making a pest of myself, gaining far more from the experience than they ever gained from me.

Finally I started winning grants to support my trips to the US. I'd spend time at Johnson Space Center in Houston or at the Cape in Florida. I'd go whenever I could and do whatever they had for me to do. And the more I learnt about human space flight, the more ludicrous it appeared.

I used to feel guilty about this double life spent shuttling between Houston and the hospital wards. I'd leave a punishing on-call stretch in a UK hospital to fly across the Atlantic and sit in on a meeting where people talked, straight-faced, about how NASA might best get people to Mars.

Later, though, as I specialised in intensive care and began to understand what we were up against in trying to treat the extremes of illness and disease, I started to wonder which was the more ridiculous pursuit: standing at the end of a bed, tilting at the windmills of critical illness, or staring down telescopes at destinations that lay waiting to be explored.

And as I learnt more about how we arrived at this point, with such inflated expectations of life and survival, I began to realise that while our medical pioneers weren't concerned with geographical conquest, they were very much in the business of exploration.

~

Extremes is a book about life: its fragility, its fractal beauty and its resilience. It is about a century during which our expectations of life transformed beyond all recognition. A time in which we took what was routinely fatal and made it survivable.

Little more than a hundred years ago, maps of the world still boasted white space: places where no human had ever trodden. But within a few short decades, these – the most hostile of the world's environments – had all been conquered. The South Pole, our mountain-tops, deep-sea trenches and the endless skies were breached in quick succession. Even the new ocean of space was finally sailed.

But it was an era that also saw medicine advance in huge leaps and bounds. With heart transplants, intensive care, trauma surgery and state-of-the-art life support, this exploration of the human body was no less extreme than our forays in the physical world.

The theme of rapid advance, using technology and science to surround our physiology like a cocoon, runs through all of the stories in this book. Human physiology is as fascinating as it is complex and, within the limits of a physical environment that can support it, extraordinarily resilient. But the way we've extended our reach – to the edge of the Earth and beyond into outer space – through the artificial systems we have built to augment our biology, is simply breathtaking.

It is this enhanced understanding of human physiology, and our new ability to protect it, that has in the past hundred years allowed us to extend our reach and simultaneously look within to the edges of our own lives.

This is a book about medicine, but also about exploration in its broadest sense – and about how, by probing the very limits of our biology, we may ultimately return with a better appreciation of precisely how our bodies work, of what life is, and of what it means to be human.

CHAPTER 1
ICE

5 Jan 1911: Captain Scott's ship *Terra Nova* seen from the interior of a teardrop-shaped ice cavern – one of expedition photographer Herbert Ponting's breathtaking Antarctic images

ROBERT FALCON SCOTT is dying, slowly succumbing to hypo-thermia in a tent pitched on the wastelands of the Ross Ice Shelf, full of the weary knowledge that he was not the first explorer to reach the South Pole – only the first to have lost an entire expeditionary party doing so. It is 1912. Antarctica is as inaccessible as it is fraught with risk; and that, of course, is its attraction, leading men to pit themselves and their lives against its challenges.

Having been beaten to the pole by Roald Amundsen's Norwegian expedition, Scott is now embarking on a race of a different kind: the scramble to write his letters to the next of kin of his expedition team, telling of their brilliance and honour, and taking responsibility for having led them to their deaths. Time is against him.

Scott's life is a property distributed across the many trillions of cells that comprise his body. Like all human beings, he exists in a state of tension. And by that I mean simply that nature seeks equipoise: it would like, as far as possible, for all things to be as equal as they can be.

The default state for an atom or molecule is electrical neutral-ity. Here the number of positively charged protons, in their composite nuclei, and negatively charged electrons, in orbit around them, is equal. But with a little effort atoms and molecules can be made to lose or gain one or more electrons, and in so doing lose their neutrality. This is achieved by imparting a little energy – through chemical reaction, radiation or electrical discharge. When they do this they become ions and change in nature. They become more dynamic and are capable of being influenced by, and generating, electrical or magnetic fields. In the body ions can flow across porous barriers, negative charge seeking to neutralise positive charge.

The machinery of our cells is designed to separate charged ions across cell membranes. That process of separation, of creating inequality, leaves a system out of step with the simple arrangement that physics would prefer, and creates the potential for something far more dynamic: a person.

To give a sense of what I mean, imagine a budget airline operating a plane that is only half full. Say that it's a long-haul flight, and that the airline chooses to expend a little energy in getting its cabin crew to cram all of those passengers like sardines into the front half of the plane, leaving the rear of the aircraft entirely empty. (This, I think you'll agree, is a situation with the potential for the release of a lot of pent-up energy.) Now imagine that the chief executive of the airline decides people can sit where they like, just so long as they pay him another £10 for the privilege. The passengers shout and swear a bit, but eventually most of them decide that being crammed into the front of the plane is worse than paying the money and being able to spread themselves out evenly across all those lovely empty seats. The result is a plane whose passengers are distributed more evenly, and an airline executive with some extra cash in his pocket.

What the airline does with the passengers and cash is what the body does with ions and energy. By expending energy in creating artificial inequality – in the case of the body, by pumping ions to where they don't want to be – and then harvesting and storing energy as the system attempts to return to equilibrium, you can save energy for later use.

We see this all around us in nature. In weather systems, for example, winds blow from areas of high pressure to those of lower pressure. They are a manifestation of inequalities in pressure, and the system's natural tendency to smooth those differences out. And in the same way that this difference leads to a wind whose energy can be harvested by turbines, so the flow of ions across cell membranes can be exploited by the human body.

The flow of ions, along with the beautifully elegant machinery that exploits it, is what makes complex life possible, what keeps the whole that is greater than the sum of its parts – the whole that is ultimately Scott – ticking over.

At medical school I failed to appreciate the importance or beauty of that biochemistry. Confronted by blackboards full of arcane symbols and equations, I chose instead to stretch out across the back row. I remember dozing gently while professors of biochemistry laboured, largely in vain, to convey an understanding of the intricacies of cellular processes: molecular pumps moving ions across cell membranes to create that all-essential inequality. In the gloom of the lecture theatre those chemical events seemed esoteric, only vaguely connected with the stuff of medicine and life. To me, astrophysicist turned medical student, it came a poor fourth in the league table of Important Stuff, after anatomy, whole body physiology and my own need for sleep.

It has taken me most of my medical career to finally appreciate the tiny processes that enable biological systems to store and release energy. These biochemical events individually appear to bear little relation to the wonder of life, when in fact collectively they *are* life; they are everything we do, everything we are.

So, to reiterate: the privilege of the human body's complexity is bought at a price: it must expend energy pumping those ions to where they don't want to be in order to keep the wheels turning. When that price is no longer affordable, simplicity reigns once again. And here simplicity is synonymous with death.

~

The environment outside the tent abhors Scott's complexity. There is more at work here than temperatures that can freeze exposed flesh in seconds. First, there is Antarctica's aridity. The continent's great sheets of ice hold water locked away, but less than a single inch of rain falls there each year. So the Ross Ice Shelf is considered a desert. Then there is its elevation: with much of the continent thrust two miles above sea level, Scott is high enough to make heavy exertion uncomfortable, even for the acclimatised. And that's not to mention the scouring Antarctic winds. All told, Antarctica is a continent of fierce extremes: the coldest, the highest, the most parched. Its climate has made it uninhabitable for all but the last hundred years of human history.

Bleak though it may be, it's important to consider how Scott's body reacts to his plummeting temperature, because understanding that process is the key to an extraordinary advance in future medical technology.

As Scott's core temperature drops, the pumps that move ions across his cell membranes are grinding slowly but surely to a halt. The process is inexorable. In the absence of energy, energy borrowed from the fuel of food and burned in the fire of the oxygen that we breathe, the pumps wind down and eventually stop. The ions begin to assume equal concentrations on either side of the cell membranes. This simple symmetry is how death begins.

Scott isn't yet ready to die. His physiology, ignorant of his predicament, is designed to battle for him, to buy him every moment that it can, to give him his best chance of survival. As Scott writes he feels the heat draining out of his hand. The blood vessels that run in his body's periphery, carrying hot blood to his skin's surface and losing that heat uselessly to the outside world, are constricting. His body hair stands on end in an effort to trap more air close to his skin. Both of these measures are an effort to reduce conductive heat losses. In the context of the Antarctic environment, however, this physiological strategy is next to useless.

Next, Scott will begin to shiver uncontrollably, generating enough heat to slow the drop in his temperature. This shivering is more than the casual tremor we might experience at a bus stop in midwinter; Scott's muscles will shake themselves as hard as they can, consuming fat and carbohydrates ravenously. This type of shivering, a last desperate attempt at staving off death, becomes an act of physical endurance in itself. It can account for fully 40% of the body's maximum exercise capacity and it will continue while there is fuel enough to do so. But shivering, no matter how athletically, is merely a holding measure – the body's method of buying time in the hope that something in its external environment will change for the better – not a solution in its own right.

As it proceeds, the deep hypothermia will go on to alter Scott's mind, making him irritable and possibly irrational. When his body's reserves of fuel run out, the shivering will stop – a respite that will only accelerate the rate at which he cools. Like a marathon runner hitting the wall, Scott is at the end of all of his reserves. There is nothing left to draw upon. Mercifully, something that looks like sleep will follow as the electrical activity in his brain begins to fail. He will slip into a coma well before the channels in the cell membranes of his heart muscle, the gatekeepers of electrical stability in that organ, find themselves compromised. Frenzied, anarchic rhythms may follow, the heart writhing uselessly like a bag of worms before finally coming to a standstill.

With his heart no longer beating, his body will be starved of its fresh supply of oxygen. But at such low temperatures the rate at which Scott's cells fail and die will be dragged out in time. The normal window of a few hundred seconds when his brain is dying and yet still alive, in which his circulation might usefully be re-established, will instead stretch to many minutes. This fact will become crucial to medical practitioners in the years ahead, as we shall see.

But for Scott there is no rescue. The seconds become minutes; the minutes hours. Scott, once a blazing furnace of life on the sub-zero wasteland of Antarctica, is now no more energetic than the ice and snow that surround him.

~

Like all living beings, we fight against the laws that govern inanimate objects in an effort to avoid equilibrium with the physical world. Through the act of living we maintain a level of complexity otherwise unknown in the universe: the ability to grow, to adapt, to reproduce, and above all, as humans, the capacity for sentience and self-awareness. It's worth stressing that as fascinating and enigmatic as neutron stars and supernovae might seem, your brain is more complicated and more impenetrable to science than either. What makes us different, what sets us apart from the inanimate matter about us, is our ability to defy entropy, to avoid the thermodynamic reorganisation which would see us reduced to a simpler lifeless state. And as the decades pass, we – the human race – become better at it, and expand the envelope in which life is possible. For all its personal tragedy, Scott's death also contains some hints about the directions in which the envelope would expand in the century that followed his doomed expedition.

Trying to conquer Antarctica forced us to embrace and understand cold and the havoc that it might wreak upon the human body. Deepening that understanding is what allowed us to continue our explorations there. And as the decades passed and our knowledge grew, we were able to overcome hypothermia. Today that understanding allows us to do far more than persist in these environments: hypothermia has become an asset to medicine, a tool for cheating death.

~

Nearly a century after Scott's expedition, a twenty-nine-year-old woman, out skiing in the mountains of Norway, suffered an accident and went through the same sequence of physiological events. She found herself as lifeless as Scott, hundreds of miles from help, trapped by ice, with her heart at a standstill as seconds became minutes, and minutes became hours. But her story has one crucial difference: she survived.

In May 1999, three junior doctors, Anna Bågenholm, Torvind Næsheim and Marie Falkenberg, were out skiing off-piste in the Kjolen Mountains of Northern Norway, near the town of Narvik. It was a beautiful evening, one of the first days of eternal sunshine at the start of the Arctic summer, and the skiing had been good. They found themselves descending into a shaded gully called the Morkhala, a place they knew well and one that boasted a good covering of snow even this late in the season. All three were expert skiers and Anna began her run confidently.

But during the descent, Anna unexpectedly lost control. Torvind and Marie watched from afar as she tumbled headlong onto a thick layer of ice covering a mountain stream. Anna slid across it on her back, and then fell through a hole into the water. Her head and her chest became trapped beneath the frozen surface. Her clothes began to soak, their extra weight carrying her deeper, dragging her downstream with the current and further beneath the ice.

Torvind and Marie arrived at the spot just in time to grab her ski boots, stopping her from vanishing under the lip of the ice. Anna was lying face up with her mouth and nose out of the water, in an air pocket. She continued to struggle, freezing, in the Arctic stream.

None of the three could have been in any doubt about the seriousness of the situation. Anna was trapped, her clothes soaked with ice-cold water, the stream carrying heat away from her body. Even in those first minutes her core temperature was beginning to plunge. Torvind called for help on his mobile phone, explaining the life-and-death predicament to the dispatcher. As doctors, Torvind, Anna and Marie had many friends and colleagues in the rescue services – the dispatcher among them. Firm in the faith that they would make every effort to expedite an emergency rescue helicopter or a mountain rescue team, Torvind returned to help with keeping Anna from slipping under the ice.

But, after what seemed to Torvind like an interminable age of waiting, he rang the dispatcher again, this time demanding to know why nobody had yet arrived. 'Yes, Torvind,' came the reply, 'we are trying as hard as we can but you must understand it takes more than three minutes to make these things happen.' To Torvind, fighting alongside Marie for Anna's life, three minutes had seemed like eternity enough.

Two rescue teams were sent; one from the top of the mountain on skis and another from the town of Narvik at its base. The ski team, led by Ketil Singstad, were the first to arrive, but they were lightly equipped and their snow shovel wasn't enough to break through the thick covering of ice. All they could do was lash a rope around Anna's feet to help Marie and Torvind stop her slipping further beneath the ice.

A Sea King helicopter had also been scrambled, but even travelling at over a hundred miles an hour it would take more than sixty minutes to reach them, and would take at least as long again to fly back to the nearest major hospital in Tromso.

Forty minutes after first becoming trapped, Anna's desperate thrashing stopped and her body went limp. The hypothermia, now profound enough to anaesthetise her brain, would soon stop her heart.

Another forty minutes passed before rescuers from the bottom of the mountain arrived, carrying with them a more substantial shovel, with a pointed tip that was finally able to break through the covering of ice.

Singstad, leading the mountain rescue team, was already deeply pessimistic, believing that their efforts now could only succeed in retrieving the body of a dead friend. Eighty minutes had passed since Anna had first fallen into the water and her body was pulled clear of the stream limp and blue. She had stopped breathing and was without a pulse.

We call what follows 'downtime' – the period from the moment of cardiac arrest until the point at which spontaneous circulation and breathing can be restored. In that interval, the process of dying begins.

Before that comes the 'crash'. The term is apt. If your physiology has crashed, the processes that keep you alive have stopped working. When confronted with a patient in cardiac arrest you, as a doctor, are staring at the wreckage of an individual, hoping desperately that something can be salvaged from the chaos. Frankly, it's a terrifying feeling.

In any accident and emergency department, anyone suffering cardiac arrest who arrives with more than a few minutes of downtime almost invariably dies or is permanently disabled. My time as a newly qualified doctor is peppered with memories of pounding down hospital corridors in the middle of the night answering the crash call: that terrifying screech from your pager accompanied by a burst of static and a voice telling you where you instantly needed to be. The experience was always grim. Of the many thousands of people who suffer cardiac arrest each year, only a handful survive to leave hospital. The odds always appeared so stacked against us, and the outcomes so poor, that over time I became deeply pessimistic about the crash calls. I remember a registrar, seeing my distress at the end of yet another failed resuscitation, putting a comforting arm around me. 'It's not really resuscitation, you know,' he said. 'It's just a funny dance we do around the dying.'

So, as the resuscitation effort began on Anna's body in the shadow of those Norwegian mountains, the challenge that she faced looked insurmountable. She had already been without a pulse for far longer than any of the patients I'd ever rushed to attend on hospital wards. And her core temperature was now perhaps more than twenty degrees lower than it should be.

Torvind insisted that they continue their resuscitation attempts. Just before 8 p.m., more than an hour and a half after first falling into the stream, Anna was winched onto the Sea King. Aboard the helicopter, moving at speed across the Norwegian landscape, the struggle to save Anna's life became a desperate scramble. The art of resuscitation, if you can call it that, is difficult even under ideal circumstances. Helicopters, with their cramped conditions, deafening noise and vibration, are among the most difficult places in which to try and work.

Once, when transferring an unstable, critically ill patient by air, I asked the pilot what the aircraft protocols were should the patient need resuscitating mid-flight. 'Just mind the doors,' he said. 'It's usually bad if you fall out.'

The key to good resuscitation is to keep the blood supplied with oxygen and moving round the body. This we achieve by breathing for the patient, ventilating them artificially – literally pumping oxygen into their lungs – and then compressing the chest rhythmically to provide something approximating a circulation. None of this is anything like as efficient or effective as the body's native heartbeat and breathing, but it buys time. In principle, it sounds pretty straightforward; in practice, there is perhaps nothing that adequately describes the sickening, repetitive crunch of ribs beneath the heel of your hand or the rising sense of desperation you feel as the minutes tick by.

∽

When they touched down at Tromso University Hospital, Anna's heart had not beaten for at least two hours. Her core temperature was measured at 13.7°C – twenty-three degrees below her normal core, and lower at that point than any surviving patient in recorded medical history. This was genuine *terra incognita*. Any attempt to resuscitate Anna further could only proceed in the knowledge that in similar situations in the past medical teams had always failed.

It is often hard to know how to act in the best interests of your patient, even when they can talk to you and tell you what they want. In the midst of resuscitation, faced with an unconscious, dying patient, you have to try to imagine what the person in front of you would say if they were able to express themselves. It is a horribly difficult call to make. Your instinct as a human being is to carry on for as long as there's a chance of survival, however slim. But your thoughts as a medical professional are different; there are harsh realities to face. Under ordinary circumstances the prognosis is horribly bleak. Even those whose hearts are successfully resuscitated run the risk of permanent and disabling damage to their brains through oxygen starvation.

But the team at Tromso decided to continue. Despite the huge amount of time that had passed since Anna's heart had first stopped, there was still the glimmer of hope that the terrible cold might also have protected and preserved her brain.

Mads Gilbert, the anaesthetist leading the resuscitation effort, moved Anna directly to the operating theatre. He knew that raising her temperature at this point was going to be a massive challenge. Warm blankets and heated rooms alone wouldn't be anything like enough. Raising the whole body through those twenty-three missing degrees would take an enormous amount of energy – equivalent to the boiling of dozens of kettles of water. To do this quickly and without harming Anna in the process, Mads knew she would have to be established on a heart–lung bypass machine, the sort of device normally reserved for open heart surgery. By removing her chilled blood, circulating it in the bypass machine, heating it and then returning it to Anna's lifeless body, her core temperature could be raised rapidly. At least that was the theory.

They wasted no time. Thirty minutes after being established on the heart–lung bypass machine, Anna's core temperature had more than doubled, reaching 31°C. The heart itself, its molecular machinery now warm enough to work again, stuttered at first, unable to regain its own essential rhythm. But eventually electricity once again began to flow through the muscle of her heart, and this was followed by waves of contraction.

At around 4 p.m. it started to beat independently for the first time in at least three hours. That first explosive beat was captured on film as an echocardiogram.

The fight was far from over. During the resuscitation the team had to place a central line: a thin tube that would enter a major blood vessel allowing them to give fluid and drugs more easily. To do this they first had to pass a needle into her chest, aiming for a target vein whose diameter was no more than a fraction of a centimetre. It is a tricky feat to pull off at the best of times. For navigation you rely upon your knowledge of anatomy and a steady hand. To make matters worse, lying next to that vein is a large pulsating artery which, as they tell you in medical school with a wry smile, is always best avoided.

But during the scrabble to save Anna's life, the team had damaged an artery sitting just behind her collarbone, on the right side of her chest. And here again the cold conspired to kill her. The bleeding that followed as a result was made far worse by Anna's hypothermic state: blood loses much of its ability to clot at low temperatures. Having laboured so hard to save her life, the team now faced the possibility that she would bleed to death.

They transfused blood, platelets and clotting factors in an effort to replace what had been lost and encourage her blood to coagulate once more. Cardiothoracic surgeons then decided to open her chest, finally allowing them to isolate the bleeding artery and stop the haemorrhage. After hours of work by dozens of people, Anna was finally stable enough to be transferred to the intensive-care unit.

Once there her lungs failed and, unable to maintain the levels of oxygen in her bloodstream, the team were forced to take the drastic step of establishing her on a device that could oxygenate her blood outside her body, functioning like a bypass circuit for her lungs. Her kidneys also failed and their function too was replaced artificially by yet another machine.

Miraculously, Anna survived even this, opening her eyes for the first time after just twelve days. But she found herself paralysed from the neck down, waking alive but quadriplegic. Later she grew angry, asking the doctors at Tromso why they had been so determined to keep her alive. Together the costs of her helicopter rescue, resuscitation and admission to the intensive-care unit added up to many tens of thousands of pounds. All of this for a woman who awoke alive but with a body that no longer appeared to work: the best that anyone might have dared hope for given how cold she'd been and how long she'd gone without a pulse. Had their endeavours truly been worth it? Should they have proceeded with the resuscitation at all?

But there is an epilogue to Anna's story. Her paralysed body did not remain that way. It wasn't an irreversible injury to her spinal cord that had left her unable to move, as is so often the case after traumatic injuries. It was instead her peripheral nerves, damaged by the extremes of cold, which had failed. And slowly but surely these nerves and her flaccid muscles began to recover and regain their function.

The nerves recovered most slowly in her extremities. Initially she could not use her arms and legs at all. And though after six weeks she was ready for discharge from the hospital, she could not go home. Anna spent another four months in a rehabilitation unit, slowly growing in strength and learning how to move once more. It was a slow process but eventually she was able to go home. Medicine had brought her this far and where it stopped her determination had to take over.

It would ultimately take six hard years of rehabilitation in all, but eventually Anna was well enough to ski again; well enough to return to complete her training as a doctor. Eventually she specialised in radiology in Tromso, in the hospital that had dared to save her life.

Anna Bågenholm is an extraordinary survivor. Her profound hypothermia was exploited by doctors to successfully resuscitate her against seemingly impossible odds. And while her survival occurred in the context of an accident, others have benefited from hypothermia by design.

∾

Esmail Dezhbod's symptoms had begun to worry him. He felt pressure in his chest and, at times, great pain. Visiting the doctor did nothing to allay those fears. After asking him some questions, his doctor gave him a physical examination and ordered a CT scan to investigate the structures within his chest. The pictures didn't lie: Esmail was in trouble. He had developed an aneurysm of his thoracic aorta, a swelling of the main arterial tributary leading from his heart. Normally no more than 3cm in diameter, this vessel had more than doubled in size, to the width of a can of Coke. And with this swelling came the risk of rupture. The greater the diameter of the vessel, the greater the risk that its wall might suddenly tear. And if that should occur the consequences would be catastrophic. The implication was clear to Esmail: he had a time bomb in his chest that might go off at any moment. Aneurysms elsewhere in the body can usually be repaired with relative ease. But in this location, so close to the heart itself, there are no easy options. The thoracic aorta carries blood from the heart into the upper body supplying, among other things, the brain. To repair it the flow would have to be interrupted by stopping the heart. And at normal body temperatures this, and the accompanying oxygen starvation, would damage the brain leading to permanent disability or death within three or four minutes.

And yet if Esmail was to survive, the repair had to be carried out. His surgeon, the leading cardiac specialist John Elefteriades, decided to carry out the procedure under conditions of deep hypothermic arrest. This entailed using a heart–lung bypass machine to cool the body to a mere 18°C before stopping the heart completely. Then, while the heart and circulation were at a standstill, Dr Elefteriades would perform the complicated repair, racing against the clock while his patient lay dying on the operating table.

~

On the day of the operation I was there to watch this remarkable feat of surgery. Though Dr Elefteriades is an old hand with the technique of deep hypothermic arrest, every time he does it it feels like a leap of faith. Once the circulation has come to a standstill he has no more than about forty-five minutes to complete the repair before irreversible damage to the patient's brain occurs. Without the induced hypothermia, he'd have just four.

Standing in the operating room, marking the moment at which Esmail's circulation comes to a stop, is a sobering experience. At this point nothing is supporting him: no drugs, no machines, no bypass circuit. Esmail's physiology is crashing in slow motion. Up until now, the surgery has proceeded in a relaxed fashion. Knife in hand, paring away the tissues around the heart, John has chatted away as if he's doing nothing more taxing than driving me to the supermarket. That demeanour changes at the moment of circulatory arrest. Now there's no time for small talk.

The hands of the clock on the wall swing round; the digital timer counts off the minutes and seconds. John lays down the stitches, elegantly and efficiently, making every movement count. He has to cut out the diseased section of aorta, a length of around 15cm or so, and then replace it with an artificial graft. To this he must stitch other tributaries supplying the brain and upper body. And all the while Esmail is dying.

The electrical activity in Esmail's brain is, at this point, unmeasurable. He is not breathing and has no pulse. Physically and biochemically he is indistinguishable from someone who is dead. And it seems impossible to believe that he might be successfully resuscitated from this state and go on to be the man he was before.

And yet, after thirty-two minutes, the repair is complete and Dr Elefteriades is ready to re-establish Esmail's circulation.

The team warms his freezing body and very quickly his heart explodes back into life, pumping beautifully, delivering a fresh supply of oxygen to Esmail's brain for the first time in over half an hour.

A day later I visit Esmail on the intensive-care unit. He is awake and well, even if he's in a little pain, and his wife stands by his bed, overjoyed to have him back.

It seems nothing short of unbelievable that to cure this man his surgeons have had to come close to killing him – using profound hypothermia to buy his survival. But Esmail is living proof that physical extremes can cure as well as kill.

∼

There is another extreme that we have recently begun to explore, defined not by environmental conditions, as in Scott's Antarctica, but by disease and injury. It's a destination that advances in medical science have taken us to. Intensive-care medicine hangs ordinary people out there at the very limits of survival, to endure perilous derangements in physiology, with the expectation that they might survive and go on to lead normal lives.

Evolution did not prepare us for life at the extremes. Only engineering and technology allow us to cheat our environment and our biological fate – and then only temporarily. One of the questions this book will address is whether technology emboldens us before we understand its consequences. Think again of that medical team heating Anna Bågenholm's chilled blood and pumping it back into her body, with only the slimmest hope that she would survive to lead a normal life. Perhaps we have no business pushing the envelope after all. Perhaps we have finally gone too far.

But Anna did make a complete recovery. As a radiologist, she's now working at the very hospital that was so determined to save her life. She owes that life to science, technology and medicine, and an understanding of the biology of deep hypothermia that is as young as she is. So goes the story of our exploration of the extreme tolerances of the human body: one of tragic loss and outrageous survival, of questions about life and death and an attempt to understand what lies in between.

~

In the end, Scott's exploration aboard the *Terra Nova* wasn't in vain either. The expedition that he led wasn't the first to reach the South Pole, but it was one with an important scientific legacy. It laid the foundations for the discipline of glaciology and found fossil specimens that would later point to an incredible truth: that the southern continents of the world had once been linked together as a single landmass. The penguin skins collected by Cherry-Garrard, Wilson and Bowers on their infamous 'worst journey in the world' would provide a benchmark sample which would later help scientists establish the persistence and bio-concentration of DDT insecticide after its introduction into the global food chain in the twentieth century.

In contrast to Scott's effort, Amundsen's team was merely in it for the race. Scott's exploits aboard the *Terra Nova* were to have been the crowning glory of a triumvirate of expeditions that included Scott's first voyage to Antarctica aboard the *Discovery* in 1902 and Shackleton's *Nimrod* expedition in 1907 – endeavours that were instrumental in opening up the continent of Antarctica to science. Scott may have died, but what he and his expedition team started at the turn of the twentieth century would in time become a wider programme of scientific research and one of fundamental importance.

By the middle of the century, the scientific survey teams of several nations had established a plethora of permanently manned bases in Antarctica. In 1985, observations by the British Antarctic Survey detected the thinning of the Earth's ozone layer around the South Polar region: the so-called 'ozone hole'. Ozone in our atmosphere absorbs ultraviolet radiation protecting us from its harmful effects. This discovery and the later realisation that ozone depletion was being catalysed by halogen atoms in chlorofluorocarbons (CFCs) led to an international ban on these substances. And by the end of the century it was these multinational scientific efforts in Antarctica that delivered some of the most convincing evidence that global warming was a real phenomenon. Scott's race to the South Pole began as an exploratory effort into the unknown which he paid for with his life and the lives of his core team. However, the legacy of Scott's exploration is discoveries that might one day save our entire planet.

~

We are walking, Torvind, Anna and I, along Mortimer Street; they have been lecturing about their experiences to an audience of doctors at the Royal Society of Medicine in London. They feel that this reliving of the past and retelling of the story is important, to change people's practice and expectations in the face of such extreme hypothermia.

There is a question I must ask Anna. It centres around the decision that we as doctors make when the odds look stacked against us. If she'd had the choice at the time of her resuscitation at Tromso – given the extraordinarily long period for which her heart had been arrested and knowing that the overwhelmingly likely outcome would have been death or a lifetime of disability – would she too have chosen to let the team proceed?

'Yes,' she tells me after a short pause. 'Because you never know.'

We carry on our stroll through London. There is a point in the pavement where water is rushing from what looks like a burst water main, flowing across the paving stones. Anna quickens her pace, breaking into a jog. For a moment I wonder if this is one of those unexpected aversions that develop after a traumatic event. She was, after all, trapped under the ice, sinking into running water. Torvind says nothing. Perhaps it's something he's seen before. I am intrigued, briefly horrified, that this might be a sign of vulnerability or weakness; perhaps the only sign that she will show.

I am still pondering this when a taxi runs through the sizeable puddle that has collected in the gutter, drenching my feet and trousers. And I realise that, after having been entombed and frozen in an Arctic stream, having endured the lowest recorded temperature of any cardiac arrest survivor in medical history, the only reason Anna's running is because she's smarter than me.

CHAPTER 2
FIRE

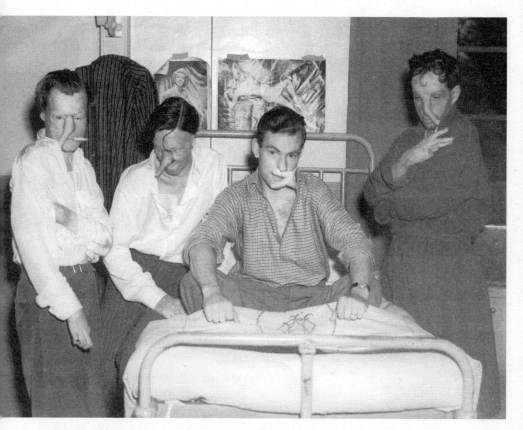

Second World War burns victims at the Queen Victoria Hospital in East Grinstead. The strips of flesh used for surgical reconstruction of their faces were kept alive by anchoring them to blood supplies at two locations. For these men the skin was bridged to the face from the forearm or chest

'HOW MUCH OF that has to go in the vehicle?' asks the pilot, pointing at the mass of tubes, cables, monitors and machines that surround my patient.

'All of it,' I tell him. The intensive-care side-room looks cramped. Its contents must somehow be transferred up to the roof of the hospital and crammed into the back of a medical evacuation helicopter.

The pilot sucks air through his teeth, doing the mental arithmetic. It's a warm day, the air is thin. His engines can only generate so much lift. The more weight we have to carry, the shorter the range of the helicopter and the more hazardous the take-off.

'How much has to be in the cabin with us?' he asks.

'All of it,' I repeat.

More air-sucking sounds.

'How much do you weigh?' he asks, looking me up and down.

'About 70kg,' I tell him.

'How about her?' he says, nodding indelicately at the nurse.

The man in the bed before us has been caught in a house fire and is badly burned. The trauma team estimates that perhaps the full thickness of skin over as much as 50% of his body has been destroyed by fire, though it's hard to be sure. Underneath the char and the blisters it is difficult to know what remains viable. Time is ticking by. Keeping him stable has taken all our efforts and we are at the end of what we can offer here in this general intensive-care ward. To give him the best chance of survival he needs to be moved to a specialist burns unit.

We pour fluid into his veins, trying to keep up with the massive evaporative losses caused by the absence of skin cover. The protein is leaking from his vessels into his tissues; the osmotic pressure is taking more fluid with it. The alveoli of his lungs are filling as a protein rich slush exudes into them. Those tiny, all-important sacs – which allow air to be brought into contact with blood, and oxygen to be exchanged for carbon dioxide – are becoming waterlogged. Things will get worse before they get better. We must move now. Tomorrow he may be too sick. But the closest specialist bed is more than two hundred miles away – too far for a safe road transfer. We'll need an aircraft. That's why the helicopter pilot is here, still doing sums in his head; weighing benefit against risk, as the patient slowly drowns in his own juices.

~

We do not immediately think of skin as an organ in the conventional sense. It lacks the solidity and the discrete locus of the more familiar viscera. That elastic but porous covering stretched over your frame, folds of flesh and imperfections that you know well enough to take for granted, fulfils an essential task. It is no less important to your continued survival than a heart or a pair of lungs.

It is tempting though to think of skin as though it were simply a barrier: a line of separation between you and the world outside; a convenient way of preventing your viscera and tissues from sloshing about in an unsightly way.

It does indeed protect, but that alone, as a description of purpose and function, is the grossest of understatements. It does much more than act as a bulwark against the abrasive world outside: it is the first line of defence against the microbial hoards massing on its surface; it prevents the excessive evaporation of the body's precious fluids; it harbours an exquisitely sensitive array of detectors that warn us of harm, allowing us to respond fast enough to avoid further injury; and it thermo-regulates – striving to keep us warm when it is cold or cool us down when it is hot.

Skin is deeper than you think; in some areas of the body it is up to half a centimetre thick. The stuff at the top is dead, a keratinised layer that serves the purpose of physical protection. Below that is living tissue that needs nutrients and a blood supply and is vulnerable to attack and injury.

If you take a micro-fine slice vertically through skin and examine it under a microscope, you can see the cellular structure of its layers. The specimens need first to be stained, otherwise the cells are largely clear and colourless. This they call histology: the study of the microscopic anatomy of cells and tissues.

In my first year at medical school, I spent many hours staring down microscopes trying to make sense of what appeared to be little more than washes of pink and purple abstract art. At the end of the course we were supposed to be able to identify any number of organs and tissues from their microscopic appearance alone. It was bewildering, like trying to identify countries from close-up, unlabelled photographs of their fields and pavements. Sitting in long rows along laboratory benches stacked with microscopes, we strained our eyes to link the lecturer's elegant verbal descriptions to the purple mess visible through the eyepiece. Some of the slide material was pink with wavy strands, like streaky bacon, and we were assured that this was a perfectly acceptable descriptive term for that tissue. In time, and with a little help, the underlying structure revealed itself and began to make sense – though not to everybody. One of my colleagues famously failed the first-year, one-hour histology exam having only written the desperate words: 'It *all* looks like bacon to me!'

When you finally get your eye in you can see that the skin is organised into distinct strata. The topmost layer, the epidermis, forms the tough barrier with which we feel so familiar. The cells of the epidermis are densely packed and further subdivided into layers. The base layer comprises stem cells that boast large purple-stained nuclei. These cells mature, eventually losing their nuclei and acquiring filaments of keratin making them more rigid. As they develop, they ascend through the epidermal layer towards the surface, finishing at the top to form a tough protective layer of dead cells.

All of that tends to reinforce our image of the epidermis as a durable but passive barrier to the outside world. Yet it is anything but passive. The layers of epidermal cells, constantly being born and marching forward, are like a never-ending conveyor belt of foot soldiers throwing themselves at the wire. The defence that they mount is spirited: they create a dry and acidic environment hostile to bacterial growth; they incorporate immune cells with tentacle-like appendages that seek out and destroy foreign bacterial cells; and they secrete enzymes and fatty chemicals to further deter would-be colonists. The fight at the surface is fierce – a defence against perpetual mechanical, chemical and biological attack. And consequently the rate of attrition among these cells is high. For a single epidermal cell, that journey – from birth in the basal layer to combat maturity on the surface of the epidermis – takes something like six weeks. The rate of replacement must match the rate of loss and the entire epidermal layer turns over anew every forty-eight days.

But the epidermis, the layer that we casually refer to as 'our skin', represents only what we can see. It is underpinned by the dermis, and it is here that the real interest lies. The epidermal layer is relatively uniform in appearance: stacks of purple-staining polyhedral cells topped by a paler weave of pink. Beneath this there is the dermis, which under the microscope looks like a vertical section through a chaotically planted vegetable garden. There are microscopic structures here that look like the cut surfaces of onions and other legumes. That infamous bacon-like connective tissue is found here, dotted with strange-looking whorls, blood vessels and tubes. This is where the skin becomes more recognisable as an organ, run through with a network of glands and vessels and studded with organelles. It is from this layer than the skin derives both its elasticity and its supply of blood and nourishment.

Together the epidermis and dermis form a waterproof but breathable layer. They have pores that are small enough to prevent ingress of water droplets but large enough that they let molecules of water vapour out. Gore-Tex clothing attempts to do the same thing, but as a breathable and waterproof barrier it achieves only the very palest imitation of skin.

But it is the sensory array that is perhaps the skin's most remarkable feature. Able to resolve point contacts little more than a millimetre apart, it's capable not only of registering heat and cold but also of differentiating between a lover's caress and the pain of a needle tip. Your skin is honed to provide a series of ever-changing inputs in response to the cruel world outside, and these shape your behaviour in such a fundamental way that you are barely aware of the process.

That holiday in the sun that you seek, the sensation of warmth on your skin, is in part a product of the pattern of receptors that activate in response to incident radiation. Think of summer; think of winter; think of plunging headlong into a pool of water. Chances are that the first thing that enters your mind is that inexpressible pattern of receptor activation that counts for warmth, cold or wetness.

The clothes that you are wearing right now are in part chosen because of the way they feel on your body. The receptors in your skin help you decide to move away from draughts, cause you to retreat from the roar of the fire or urge you to get out of a chilly pool of water. No sight or smell or sound could compel you to behave quite so urgently.

Consider then this finely tuned early-warning system linked to a consciousness that understands both luxurious pleasure and intense pain. Then imagine setting it on fire.

~

Though painful, superficial burns involving only the epidermis are little more than that. Reddened by the dilation of blood vessels beneath and the inflammation of the tissues, they are rapidly healed and restored by the perpetual marching of those regenerating epidermal cells.

Even burns which extend below, into the upper two-thirds of the dermis, retain the ability to heal and cover with new skin. A patch injured in this way generates islands of new epidermal cells that spread and eventually coalesce, replacing what was lost. These burns leave the bulk of the sensory architecture intact and are exquisitely painful. Damage to the tissue around pain receptors leaves them constantly firing. Inflammation – the process that marshals cells of the body's defences to fight infection and deal with injury – retunes the pain receptors, making them hypersensitive. The same process brings fluid into the wound, producing the blistered, weeping appearance, separating healthy tissue from that which is dead or irreversibly injured.

It takes great effort to survive a serious burn. Certainly they are among the most formidable injuries to manage in emergency medicine. After determined and skilful resuscitation specialist burns treatment must follow. This care is complex. In addition to repairing and replacing damaged skin it must also accomplish the difficult task of compensating for the failure of an essential organ.

~

The man we are trying to cram into the back of that helicopter has already been the object of frenzied medical attention.

Burnt larynxes can swell and occlude; smoke inhalation can prevent the lungs from exchanging oxygen and carbon dioxide; and poisonous gases – carbon monoxide and fumes from furniture and building materials – can asphyxiate. All of these will kill victims of fires in minutes or even seconds; long before the consequences of any external burn injury can manifest themselves.

And yet, if the opening minutes of the injury can be survived, the damage to the skin leaves a formidable constellation of problems. The vapour barrier function is lost and the body's water evaporates uncontrollably from denuded body surfaces at a rate that is difficult to anticipate intuitively. The losses are both invisible and incredibly rapid. Without the cover of skin, your body dehydrates as surely as a wet sponge left out in the sun. And the burn itself triggers a severe inflammatory response in the body, compromising the integrity of the blood vessels, making them more porous and permeable, leaving them to spill fluid into surrounding tissues.

Such is the severity of this reaction to burns that at the start of the twentieth century, victims with as little as 10% to 20% full-thickness burns over the surface area of their body would often die.

Thankfully that has changed, but when it comes to burns the totality of the surface area involved remains one of the key prognostic indicators. In the care of burns we were taught a rule of thumb: the percentage area of body involved in a full-thickness burn plus the age of the patient gave the percentage chance that the patient would not survive. Under this scheme a sixty-year-old man with full-thickness burns over 40% of his body, for example, would not be expected to live. Today that is an outmoded concept; older people with more extensive burns are surviving against expectations, due in large part to the efforts of specialist burns units and the hard-won lessons of the twentieth century.

Aggressive resuscitation with fluids, trauma systems and early transfer to specialist units have all helped to improve survival rates. But back in 1940 the medical fraternity knew little or nothing of all of this.

For a pioneering generation of RAF fighter pilots, immolation was a risk they took every time they climbed into the cockpit. It's their experiences that have shaped – and continue to shape – the scope and ambition of burns treatments to this day.

31 August 1940: the Battle of Britain was reaching a critical phase. The Luftwaffe was mercilessly bombing the Royal Air Force's airfields. Fighter Command was losing aircraft faster than they could be replaced, and their remaining pilots were fatigued. Across the South of England, fighter squadrons had scrambled time and time again to meet waves of German bombers escorted by Messerschmitt fighter planes. In the heat of that combat, the air was full of glowing munitions and criss-crossing aircraft. At RAF Kenley, on the outskirts of South London, thirty-two-year-old Tom Gleave had taken over command of 253 Squadron from Howard Morely Starr, who had been killed the previous day. Despite the ever-present danger, Gleave still found it impossible not to be captivated by the spectacle of the world as seen from the air. Scrambling from Kenley that day, he climbed quickly into a perfectly blue sky and dazzling sunlight. North of him lay the River Thames, glittering as it snaked its way through the London sprawl; to his south he could make out the Kent coastline shimmering in the summer haze. And below, unfurling at hundreds of miles an hour, rolled the patchwork quilt of the English Home Counties.

Having shot down no fewer than five Messerschmitt 109s while on patrol the day before, Gleave was in confident form. With reports of a large formation of enemy aircraft converging to attack Biggin Hill Airfield, Gleave turned with his section of three Hurricanes to assist in its defence.

Ploughing north, Gleave searched the sky for evidence of the enemy. Suddenly he found the sky above him dark with aircraft: column upon column of Junker 88 bombers. He and his section remained unseen, less than 1,000ft below and beyond them.

Keen to press the attack before the German turret gunners had a chance to fire, Gleave pulled the nose of his Hurricane up, took aim and raked the fifth bomber in the line with cannon shells. The smell of spent cartridges filled the cockpit; the Hurricane's nose dipped with the repeated recoil of the guns. Gleave pulled the control column, kicked hard on the rudder pedal, turned and dived away. Levelling out he began to climb again, attacking another bomber in the formation. For his third pass he decided he would take on the lead aircraft, which had already begun its dive in preparation for a bombing run. But before he could manoeuvre into position he heard the click of a round striking his aircraft and felt a sudden heat rising in the cockpit.

Gleave glanced down. Flames were pushing into the right side of the cockpit from below; the fuel tank buried in the root of his starboard wing was alight. He rocked the Hurricane hard, and slipped it sideways in the vain hope that this would somehow quench the fire. But the flames only grew more fierce, wrapping round his feet and climbing to reach his shoulders. Plywood and fabric burst rapidly into flames around him, accelerated by fuel from the breached tanks. In a few short seconds the centre of Gleave's cockpit had become the head of a blowtorch. The aluminium sheet in which the dials of his control panel were set began to melt. But he was far too high to ditch the aircraft; there was nothing left he could do but attempt to bail out.

Gleave was still tethered to his vehicle by the oxygen mask and radio cord attached to his helmet. He reached down to rip these from their attachments but the searing heat beat him back. With his arms outstretched he could see the skin of his hands bubbling and charring. He unclipped his harness and tried to raise himself from his seat, but could no longer find the strength. Trapped with his plane ablaze and falling from the sky, Gleave's hand fell to the butt of his service revolver and momentarily he considered a quicker, less painful end.

However, there was one last chance. If he could open the canopy, pitch the aircraft forward and flip it over onto its back, then perhaps the manoeuvre would fling him out. Gleave tore his flying helmet off, severing his last connections to the Hurricane. He slid the canopy open, shoved the control column forward and then everything around him exploded.

He found himself propelled for many yards, enveloped in a ball of flames, finally breaking free into thin air and then tumbling towards the ground. His burnt hand now reached again, not for releases or a revolver, but for the D-ring of his ripcord.

Finding it he pulled hard and felt the unfurling of his parachute and a comforting tug as its silk canopy inflated above him. The roar of his engine and the cockpit inferno had been replaced by silence and a serene view of the English countryside that oscillated gently as he swung to-and-fro beneath his parachute.

He hit the ground hard and fell onto his side, somehow managing to avoid further injury. Releasing his parachute, Gleave eventually found the strength to get to his feet. His boots and socks appeared to be intact and largely unburnt. But that was where normality ended.

His trousers had gone except for a small patch protected by the parachute harness. Above his ankle the skin over his right leg had blistered and ballooned along its whole length. His left leg was in much the same state, save for a patch of skin over his thigh which had been relatively spared. The underside of his arms and elbows were burnt and the skin hung in charred folds from his hands and wrists. His head and neck too had been exposed to the inferno and his eyes were little more than slits. His nose had been all but destroyed.

Somehow he staggered across the field towards a gate on its far side, shouting for help as he went. 'RAF pilot,' he blurted out. 'I want a doctor.'

~

You can just about bear to hang on to a mug of hot tea at 42°C. That's just five degrees higher than your normal core body temperature; pretty unimpressive really, but that is where the limits of human endurance lie. Underpinning the sensation that forces you to drop the cup is a clever receptor: a weave of proteins attached to an ion channel that controls whether it opens or closes according to how hot it is, and converts the sensation of heat into pain.

For a species so wedded to exploration, such a modest thermal tolerance seems strangely limiting. But the proteins from which that receptor is built, and those that stack together to build everything from your digestive tract to your DNA, start to fall apart at 45°C. And that's where the physiology of thermal injury starts. As temperatures climb, cells lose their capacity to self-repair, vessels begin to coagulate and tissues become irreversibly altered and later begin to die. All of this happens as you approach a temperature of around 60°C. Aircraft fuel, properly supplied with oxygen, can burn at over a thousand degrees centigrade.

~

Tom Gleave woke underneath a bed in darkness. Close by there was the crump of bombs hitting home. He was at Orpington General Hospital in the middle of an air raid; the bed his make-shift shelter. He had survived, but the surgical teams at Orpington had little experience of such severe burns. They had covered his wounds in solutions of gentian violet and tannic acid, the former for its antiseptic properties, the latter as a kind of chemical dress-ing that would cover wounded areas and then harden as a supposed protective barrier to infection. As a therapy for signifi-cant burn injury these measures were at best ineffective. Worse still, they encouraged scarring and infection. And the dressings, simple dry gauze and bandages, stuck hopelessly to Gleave's weeping wounds, pulling off skin whenever they were changed.

Inevitably sepsis set in and Gleave spent many days slipping in and out of consciousness, hallucinating and delirious with fever. But he rallied and survived this too and after several weeks the medical team at Orpington decided to transfer him to the Queen Victoria Hospital in East Grinstead, which had devel-oped a reputation for plastic reconstructive surgery under the leadership of one Archibald McIndoe.

When the orderlies arrived to prepare Gleave for the

journey, they dressed him in full military uniform, shearing dressings from delicate, partially healed layers of skin. The fact that the medical staff at Orpington allowed this reflects how little was understood about the nature and therapy of burn injuries at that time. But the hellish, seventeen-mile road trip to East Grinstead delivered Gleave to the care of McIndoe and his team and the start of his reconstruction and rehabilitation.

∾

Ward Three at Queen Victoria Hospital was a wooden-walled hut linked to the main hospital building by covered walkways. Within resided a cadre of men disfigured by fire, and in 1940 the most severely injured of these were Hurricane pilots.

To the rear of the ward was an extension which housed a bath through which a warm, weak salt solution was circulated. The bath was arranged so that a flow ran through it, exchanging a gallon a minute. Afterwards drying was achieved by standing the men naked in front of large heating lamps, thus avoiding the abrasion of towelling. The pilots came to call this 'The Spa' and, with some trepidation, it was into this tub that Gleave found himself being lowered on the evening that he arrived.

He needn't have worried. His wounds were bathed properly for the first time and old dressings floated away without pulling skin with them. Later his cleansed wounds were dressed with Vaseline-coated gauze: an invention of McIndoe's that covered the wound but stopped the dressings sticking.

A few days later, McIndoe came to Tom's bedside and explained what needed to be done. It would take many months and dozens of surgeries, McIndoe explained. 'You won't like it,' he said, 'but it'll be worth it.' Something in the manner of this surgeon, standing there peering at him through horn-rimmed glasses, gave Gleave confidence. And for the first time since the inferno, he felt as though he had been thrown a lifeline.

Archie McIndoe was a New Zealander who had travelled to the UK in 1930, with his wife and two daughters, at the suggestion of Berkeley Moynihan and with the hope that he might forge a career in abdominal surgery. But he arrived to discover that there was no job attached to Lord Moynihan's invitation and, with a young family to support, he was forced to change course. McIndoe's older cousin – and esteemed surgeon – Harold Gilles came to the rescue. Gillies found McIndoe work in his private surgical practice and eventually secured him a place at St Bartholomew's Hospital in London. Here McIndoe had got a taste for the possibilities that lay in reconstructive surgery. Harold Gilles had pioneered techniques of plastic surgery during the First World War, and the first patient to undergo this type of surgery was a sailor burnt at the Battle of Jutand. In retrospect, the cosmetic result of these surgeries was primitive at best. But at the time the idea that badly damaged faces might be reconstructed in this way was revolutionary. It would fall to McIndoe to refine and advance these techniques and for him the air war of the Battle of Britain would provide a defining challenge.

First, Gleave got new eyelids pinched from the unburnt skin on his upper left arm. These tiny islands of skin were removed and sculpted into place. They were so small they could rapidly establish themselves at their new location on Gleave's face, seizing upon the bed of vessels and perfused tissues that lay there waiting to be covered, like a minuscule sod of earth being transferred from one lawn to another. Oxygen and nutrients readily diffused into these small tokens of flesh. And the wounds left by taking these grafts were discrete enough that they could be left to heal spontaneously.

But larger patches can't be moved in this way; their needs are more demanding. In plastic surgery the battle, as Harold Gilles once put it, is between blood supply and beauty. A full thickness flap of skin of about the size of an adult's palm, cut out and moved as a single slab, will die before it has a chance to pick up a new supply of blood.

To get round this problem McIndoe would raise a flap of skin, from an unburnt area – often the abdomen – leaving it attached at one edge like a trapdoor. This kept the flap alive, supplied by vessels running through its attached edge, but left it fixed in position. He would then fold the sheet of skin into a tube, stitching its long edges to one another, to protect its raw under-surface from infection.

To move this tube of skin he would make an incision in the patient's arm and form a pocket into which its free edges could be tucked. He would then stitch the flap into place, fastening arm to abdomen in the process, and wait for it to heal into position. This procedure could take weeks, during which the patient found themselves handicapped by their strange new anatomical arrangement.

Once the flap had established itself in the pocket, its link with the abdomen could be severed. This arduous process left a flap of skin, previously from the abdomen or chest, now drawing its blood supply from the patient's arm and free to be moved to any location which the arm could reach. This process of walking a tube of skin end over end from one part of the body to another was known as waltzing. The technique had been invented by Gilles, but McIndoe brought it to maturity, waltzing flaps from larger areas than ever before. It was this that provided the plasticity in McIndoe's reconstructive technique, allowing him to address larger areas of burn injury by walking skin up from distant uninjured sites.

If it sounds like the sole goal of this macabre enterprise was the movement of slabs of skin from one area, to provide cover to another, it was not. Aesthetic considerations were at the heart of McIndoe's work. It was not enough simply to provide protective coverage; cosmesis was essential. Skin is indeed one of the principal organs through which we are able to experience the world. But McIndoe also understood that it is also the means through which the world experiences us. When the war started and the toll of burned airmen began to become apparent, it was thought that the best thing you could do with the victims was to institutionalise them away from society, with the intention of protecting one from the other. But McIndoe was unwilling to accept this fate for his patients, and his efforts in reconstructing the injured went far beyond surgical innovation. McIndoe would give them new faces, but they in turn would be expected to face the world again.

Ward 3 became famous for its feats of plastic reconstruction and notorious for the antics of its resident airmen. McIndoe resisted the militarisation of the ward. The Queen Victoria Hospital was his – quite literally. The Air Ministry had seen that control of the facility was signed over to McIndoe and it was run by his rules. Military discipline was relaxed and rank ceased to have significance among the men in the beds, except of course when it came to McIndoe, whom they referred to as 'The Maestro', 'The Boss' or simply 'Sir'. Beer kegs stood freely accessible on the ward itself and at times it came to resemble something like a cross between a bordello and a working men's club.

All of this did something to distract from the grimness of the pilots' reality. Not only were they assaulted by disturbing odours of char and infection, but they were exposed to a series of strange new procedures that left them with arms stitched temporarily to chests, abdomens and faces, initially leaving them looking more bizarre than even their injuries had.

Confronted with long drawn-out weeks of suffering, with free beer as their only real comfort, the patients of Ward 3 attempted to inject some purpose into their lives: they set up a drinking club.

At first they stumbled with the name, coming up with 'the Maxillonians' in reference to their ongoing maxillofacial surgeries. But they quickly realised that this was unwieldy and didn't quite capture the spirit of their circumstances. They were a new breed of casualty, being dealt with by a pioneer surgeon armed with groundbreaking techniques. They knew at heart that they were the subjects of experimentation – however well-intentioned. And so the drinking party reformed under a new name, 'The Guinea Pig Club', with Tom Gleave among its founding members, the first and only Chief Guinea Pig.

The club's activities moved rapidly beyond drinking and

singing around pianos to rehabilitation and support. McIndoe orchestrated trips to East Grinstead. There the soldiers were dispatched, often under protest, to mix with the local population. The people of East Grinstead grew to embrace McIndoe and his army of strangely reconstructed men. They would make every effort to accommodate them, removing mirrors from their pubs, cafés and restaurants and taking care to give the lives of McIndoe's Guinea Pigs a veneer of normality. In time East Grinstead became known as 'The Town That Never Stared', and it served as the perfect preparation for the Guinea Pigs' re-entry into a world that inevitably would.

Gallows humour became *de rigueur* for the Guinea Pigs. They recruited a treasurer with badly burnt legs so that he wouldn't run off with the petty cash, and a secretary whose fingers had been injured so he couldn't keep minutes. After the Battle of Britain the Guinea Pig Club was still tiny. But with the onset of the bombing campaign those numbers rapidly swelled and by the end its membership numbered more than six hundred. They were testing times that saw McIndoe and his team forced to refine their techniques as they went, learning from successes as well as mistakes. But these lessons would transform the field of plastic surgery.

~

The practice of military medicine during the Second World War focused principally upon the salvage of life and limb.

McIndoe didn't save the lives of the Guinea Pig Club, at least not immediately. That task was achieved by the hospitals that received them. But McIndoe's work and the experience of those he treated taught clinicians that there was something at least as precious as life that modern medicine might salvage.

Today plastic surgery has its own image problem. All too often we associate it with tummy tucks and celebrity nose jobs rather than the plight of burns victims.

But plastic surgery retains many of the values that drove McIndoe and his heroic club of Guinea Pigs. It is, in the main, still about the restoration of function and appearance to people whose lives have been cruelly and irreversibly altered by illness and injury. And the fact that we, in modern times, have been able to move beyond the pursuit of simple survival is perhaps something to celebrate.

Plasticity, in the context of surgery, refers to the ability to mould and alter the appearance of the body. McIndoe was able to find areas of healthy skin and move them to cover those which had been destroyed by fire. More than this, he was able to achieve a result that was aesthetically acceptable. But there were limits. These waltzed skin flaps were supplied by an indefinite weave of capillaries and venules running through the layers of tissue. This blood supply was tenuous and flaps of this type had to be limited in length and breadth if they were to survive. More extensive injuries were not so easily addressed using this technique.

Larger and thicker areas of skin need much greater volumes of blood flowing through them to keep them alive. In terms of blood supply it is akin to the difference between the needs of a village that subsists on the trickle of dozens of mountain streams that run through it, and those of a city built on the banks of a coursing river.

This problem could, in theory, be overcome if a block of tissue could be harvested along with the artery and vein that supplied and drained it. These vessels could then be connected to the body's core circulation at the new site to which the graft was being moved. By moving and then connecting a flap directly to the circulation in this way it could be perfused with a rich flow of blood and made viable more or less immediately.

If this could be achieved then McIndoe's waltzing flaps would no longer be necessary. Instead, free flaps of skin and tissue could be taken and moved in a single operation. No longer would the patient have to wait contorted for weeks while the tissue established a useful blood supply or be forced to undergo countless operations.

But the vessels which supply and drain such flaps of skin, though huge compared with capillary networks, are still vessels of tiny calibre and connecting them demanded a level of surgical precision previously unknown. No one could cut and stitch vessels whose diameter might be little more than a millimetre under the naked eye. For this they would need a new but familiar tool.

It turns out that microscopes have more to offer surgery than the bafflement of its students. By the 1970s, microsurgery was an established technique. The skin, whose anatomy had been so well explored by histologists with microscopes, could now be manipulated surgically using the same tool. In time the use of optical aids to magnify the view of the surgeon would become as essential to the art of plastic surgery as McIndoe's scissors or scalpel. The ability to operate under a greatly magnified field of view made finer procedures, including the connection of blood vessels and nerves, a reality. And so for the first time flaps of skin, muscle and bone could be moved *en bloc* from one location to another in a single bound – the so called 'free flap'.

This development massively expanded the plastic surgeons' repertoire and gave rise to a plethora of important and exciting new techniques. But the selection of flaps that could be used was still relatively narrow. And though the grafts made available by this method could be moved quickly and could cover much larger areas, the aesthetic result was sometimes less than satisfactory. Authorities of the time referred to these early, free flap grafts as 'hamburgers of tissue' or 'globs and blobs'.

To be of genuine value in aesthetic reconstruction, the library of skin and tissue flaps that plastic surgeons could draw upon needed to be greatly expanded. But knowledge of the vascular anatomy of skin – its relationship to the core circulation – wasn't yet at a point where this was possible.

In the 1980s Australian plastic surgeon Ian Taylor recognised this, and undertook a massive remapping of the circulation of the skin, and in so doing reconceptualised the anatomy of human skin and its relationship to the circulation.

Prior to this work, understanding of the connection between the core identifiable vessels of the circulation and the supply of more peripheral structures was poor. The body has a network of named arteries and veins which are reproducible from one individual to the next with little variation. These divide ultimately to form more variable, less distinguishable vessels. By the time they arrive at the planes of tissue underpinning the skin, the network has degenerated into a complex weave of small and largely nameless tributaries.

This was fine if you were a surgeon operating on, say, the heart or the liver, where the principal vessels are generally constant in appearance, well mapped by anatomists and immediately recognisable. But for surgeons interested in moving units of flesh and skin around, it was like having an atlas of Great Britain that included only motorways and the larger A roads and trying to navigate a route to a remote farm in the Scottish Highlands.

Taylor injected radiopaque dyes into the skin of countless cadavers and took X-ray images. In so doing he generated stunning images of the network of small but remarkably consistent vessels that connected the core circulation to the skin and tissues above.

Understanding these connections and the routes that vessels took as they rose up from deeper structures, weaving between planes of muscle and fat, allowed him to deconstruct the body into a three-dimensional jigsaw. Taylor called the pieces of the jigsaw angiosomes and together they constituted a library from which units of tissue, skin and bone could be drawn and reliably transferred to almost anywhere on the human body. But the battle between blood supply and beauty was far from over.

∽

The face derives its blood supply from a branch of the carotid artery. This divides low in the neck into a deep internal branch and one that runs more superficially. It is from the superficial division that the face gains its blood supply. From this there are branches aplenty, enough that we as doctors in training employed a variety of, mostly obscene, mnemonics to help remember them.

Run the tip of your finger gently back along the line of your jaw until the point just before it turns up towards your ear. At this point you can feel the pulse of the facial artery as it runs just below the surface of the skin.

From here it breaks over the surface of the face, with smaller vessels running above and below the lips, and branches that run alongside the nose and then up to the eyes. And this shower of arteries joins with other branches of the external carotid artery that also creep across the face. This arrangement supplies both the facial skin and well over a dozen muscles that are involved in eating and facial expression. It is the complexity of the arterial blood supply which surgeons had feared might prove an insurmountable challenge when it came to attempts at full face transplants. But more recently it was discovered that the blood vessel connections required to supply and drain the face might be fewer and simpler than previously thought. This realisation would take the concept of a full face transplant from a thing of science fiction into realms of science fact.

~

There is such a thing as life after death. It's called transplant medicine. After death a patient's heart, lungs, liver and kidneys can be donated to give the gift of life. And not just to one person – several lives can be saved by that single act of generosity. But death must come first.

Every year in the UK 1,000 people die waiting for an organ transplant. It is possible for patients to receive an organ, removed from a donor, after the heart has stopped beating. This we call non-heart-beating organ donation and it is a measure that has greatly increased the number of organs available for life-saving donations.

But waiting until the heart has stopped beating before beginning the transplant process means that the organs become deprived of a fresh supply of blood and oxygen. Once that has ceased the organs begin the process of dying and so there is a greater risk that they will fail to function properly after transplantation.

Some organs are more resilient to this than others. Kidneys in particular can endure long periods of little or no blood supply and still be resuscitated. But organs with higher metabolic demands, such as the lungs and the heart, fare less well. It is because of this that a new definition of death was coined around the time of the first heart transplants, to give surgeons the best chance of obtaining a heart that might survive the transplant process and function well.

After severe head injuries the brain can sometimes be so damaged that its higher functions are lost, leaving only the most essential reflexive processes intact. The intrinsic rhythms that drive your heart or the automated activity that drives your digestion, for example, can continue even if everything that is essentially you has ceased to be.

It is this that defines brain stem death: the irreversible and permanent loss of consciousness and cognition. It is a state as final as that which accompanies the standstill of a heart and the arrest of breathing. A heartbeat may remain and breathing might be supported artificially, giving the outward appearance of life, but the stuff that dreams are made of, the elements that define a human being, are no longer present.

When this occurs, the organs continue to be supported by the beating heart that remains, even though death has already occurred. But it is from these tragic losses, usually from accidents or massive strokes, that the best hope of new life can come, since brain death allows organs to be given in the best possible condition.

But these conversations that we have with the relatives and close friends of patients, in softly lit rooms on hospital corridors, are among the hardest in all of medical practice. For the team that approached a recently bereaved family, somewhere in New England in March 2011, to ask for their consent to donate not only a heart or a liver but also a face, the task must have seemed impossibly onerous.

They took their time, talking over the intricacies of the procedure. They told them that it was among the first of its kind in the world – and in that respect as experimental as much of McIndoe's early work. There could be no coercion, only openness.

There were, however, reassurances. The transplant team made it clear that the recipient of the donated face would not resemble their loved one. Once transplanted, the face, laid upon a new underlying structure of bone and tissue, would be as unique in appearance as any other. Neither identity nor appearance would be transferred.

But there were also difficult realities to confront. After the retrieval of a face, efforts are made to reconstruct the appearance of the donor. Casts of the face are taken and silicon masks are sometimes fashioned. But none of these restores the donor's appearance enough to allow them to lie in state, in an open casket. All of this had to be understood and accepted. After deliberation and despite the enormity of the request, the family gave their consent.

~

That day, plastic surgeon Bohdan Pomahac found himself sitting in the back of a private jet taxiing on the runway at Boston airport, waiting to take off. He was there leading a transplant team, making ready to head out and retrieve a donor organ. The plane was one of several regularly chartered by the hospital's transplant service. Hearts, lungs, livers, kidneys and other organs were often ferried urgently across the United States in this way. But this was different. That evening Pomahac was going out to retrieve an organ as a prelude to a procedure which the United States had never before seen: the transplant of a complete face.

Pomahac had waited a long time for this opportunity, and fought hard just to gain permission to attempt the operation. At the time only one other full face transplant had ever been carried out – by a team in Spain a year earlier. Pomahac was nevertheless convinced that this procedure offered the only real hope for people who had suffered catastrophic facial injuries. But not everyone was of the same mind. He petitioned the Institutional Review Board (IRB) at the Brigham and Women's Hospital repeatedly. The board, tasked with making sure that both the science and ethics of the proposed procedure were sound, were supportive but took some time to convince. The difficulty lay in the fact that, unlike other transplant surgery, the transfer of a face did not ameliorate life-threatening illness. The review board had to weigh the very real risks that the procedure carried against its perceived aesthetic benefits. And it wasn't just the surgery that might present a threat.

To be able to accept a transplant from another individual, the recipient's immune system must be heavily suppressed to stop the newly grafted organ from coming under attack. For ordinary organ transplants, the tissue type of the donor organ must be matched as closely as possible to that of the recipient. Part of the body's formidable defence against infection lies in its ability to distinguish foreign proteins and tissues from its own – a function fulfilled by the white blood cells patrolling in our circulation.

Once recognised as 'other', foreign bodies are attacked by battalions of immunological cells which damage, destroy and later engulf them. Without this defence, the simplest of infections would prove lethal. But if you want a patient to receive a transplant organ from another individual, these defences work against you. The newly grafted organ is detected, attacked and eventually rejected by the body.

During the Second World War, plastic surgeons were aware that skin grafts taken from donors related to the recipient survived longer than those taken from unrelated individuals. Precisely why this should be the case was unknown, but it gave pause for thought. Archie McIndoe himself had observed that grafts could be exchanged between identical twins without fear of rejection. Today donor and recipient are matched as closely as possible with respect to specific marker proteins expressed by their cells. These proteins are like flags on the mast of a ship at war, announcing its sovereignty and distinguishing it from the naval vessels of a hostile foreign power. For the cells of the human body, exhibiting the wrong surface marker proteins is akin to flying hostile flags and in turn provokes attack.

Matching surface proteins as closely as possible provides a degree of protection, but ultimately only delays the onset of rejection. To ensure graft survival the recipient's immune system must be suppressed, which in turn exposes the recipient to the risk of overwhelming and potentially fatal infection. Pomahac's many appeals to the IRB at the Brigham and Women's Hospital had gone some way to opening their minds, but each case would still have to be decided on its individual merits. And so Pomahac began his search, a search that would lead him to Dallas Wiens.

Pomahac first heard of Wiens at a meeting of the American Society of Plastic Surgeons in 2009. The surgeon was due to present case reports of successes that he had had with partial face transplants. But speaking before him was Dr Jeff Janis, a surgeon from Texas, who told the story of a man who had suffered the near total destruction of his face through electrocution. Dallas Wiens had been helping to paint a local church in his home town and had climbed into a cherry picker in order to reach the roof. What happened next remains unclear. As the basket containing Dallas rose from the ground, he appears to have got close enough to a high-voltage power line for it to discharge through his body. The electrocution continued for many seconds; nearly long enough to kill Dallas and more than long enough to burn and almost completely destroy his face.

Wiens was resuscitated in the emergency room of the Parkland Memorial Hospital in Dallas, Texas. The scenes would have been distressing even for seasoned healthcare professionals. Electrical burns are caused by the heating effect of the current as it passes through tissues. The resultant burns run deep and electrical involvement of the heart can lead to immediate cardiac arrest. The power line had discharged through his head, heating and then burning the full thickness of skin over his entire face. The charge running through his body cauterised his face, reducing it to a coagulated mass.

Dallas was close to death when he arrived at Parkland. The resuscitating surgeons wondered how hard a fight for life they should mount. Seeing how completely his appearance had been destroyed, they initially wondered if anyone would want to survive in such a disfigured form.

A face fulfils a role that goes well beyond appearance. Its orifices form the conduits through which air is conducted into our lungs and through which food begins its journey towards our digestive tract. It is the sole seat of three of our five senses:

sight, smell and taste. From what the resuscitating team could see, much of that had been utterly obliterated. Even if Dallas could be resuscitated, what quality of life could this man possibly hope for?

Nevertheless they continued and later Jeff Janis' plastic surgical team would cover Wiens' head by raising large, free flaps of tissue from Dallas' back and moving them up onto his face. But Janis was open about the fact that this effort was a life-saving measure whose goal was to cover and manage the wound left by the electrical burn. Even after this work had fully healed it was clear that Dallas would need a more radical solution if any meaningful reconstruction were to be realised.

In the presentation that Pomahac saw, Dallas looked as though he had been massively injured. A huge featureless graft had been pulled into place where his face had once been. Pomahac remembers thinking to himself that the man in the slides was so completely disfigured that he no longer looked human.

After they got off the stage, Janis and Pomahac got chatting. Having heard of Pomahac's pioneering work in the field of partial facial transplants, Janis suggested that perhaps Pomahac's team at the Brigham could help Dallas. But Pomahac was pessimistic; he was unsure how much of the structure of the face remained intact underneath the graft. To be reconstructed, Dallas would need a full face transplant and for this the underlying blood vessels would have to be intact. Judging by the details of the medical report and the photographs of Dallas that Pomahac had seen, he doubted that this could be the case.

Still, he decided to investigate further. Pomahac brought Dallas to Boston and began to assess him. Many aspects of his injury were at least as bad as he had feared. He was blind and had lost one eye. The structure of the nose had been entirely destroyed, he had no lips and where there should have been a mouth there was only a slit. Dallas was reduced to drinking through a straw and when he ate he had trouble keeping food in his mouth. He could just about speak but the words were sometimes muffled and difficult to comprehend.

But as Pomahac came to know Dallas better, he couldn't help but be won over by the force of his personality. Here was a patient who remained positive despite the accident, and open about the disfigurement he had suffered. He was also realistic in his expectations and clear about his motivations. The injury had left Dallas blind and so, one might assume, less conscious of his facial features. But the opposite was true. In conversations with Pomahac, Dallas explained the profound discomfort he felt in sensing the reactions of others to his appearance, the silence that fell in a previously busy restaurant when he sat down to eat and the hush that filled rooms in his presence. He was acutely conscious of all of this. But most of all he worried about how his young daughter would cope with questions and comments from friends as she grew older.

While this didn't alter Pomahac's technical decisions, it certainly shifted his emphasis. He desperately wanted to help this man; this was something that went beyond the ordinary duty of care.

Pomahac had been preparing for the possibility of performing a full face transplant for more than two years, assembling a crack team from across the medical disciplines. He knew that the surgery itself was just the centrepiece; a host of clinicians and other healthcare professionals would be necessary to make Pomahac's ambition a reality. For this plunge into the unknown he would have to make sure that his team were meticulously prepared. This responsibility he gave to his friend and colleague Tom Edrich, an anaesthetist.

By this stage the team were on call twenty-four hours a day, waiting for the phone to ring calling them to action. Meanwhile there was plenty to think about. Where should the intravenous lines be sited? What degree of immunosuppression would protect the graft from rejection without running unacceptable risks? And for Edrich there was the question of how to prevent the patient's airway from closing and suffocating him after the anaesthetic had taken effect. Normally this would be achieved by inserting a tube through the mouth and threading it into the windpipe. But with burns patients, mouths were often too badly distorted to allow this to happen.

Meanwhile the question of whether or not Dallas could be a candidate for facial transplantation depended upon the state of the blood supply that remained. Pomahac's team set about conducting an extensive mapping of his vascular anatomy, injecting liquid opaque to radiation into Dallas' veins and performing computerised tomography (CT) scans and magnetic resonance imaging (MRI), revealing the delicate network of vessels woven below.

After many weeks of assessment Pomahac's team decided that, despite the apparent damage, the key blood vessels remained intact. There was a good chance that Dallas would be able to receive a face transplant. Pomahac began working him up as a candidate, profiling his immune type so that the transplant teams could begin their search for a donor who was a suitable match. All that then remained was for them to find a donor whose immunotype was a close enough match.

They waited for several months. Then one day Edrich's phone rang. It was Pomahac and though his voice remained level, Edrich could detect more than a hint of excitement. 'We've got a face,' he said.

～

For the next two days nobody involved in the face transplant slept very much. Dallas was told to make his way from his home in Texas to the Brigham and Women's Hospital in Boston as quickly as he could. Meanwhile, Pomahac set out aboard a jet aircraft to retrieve the face from its donor. He was not the only transplant surgeon on the retrieval mission. Under ordinary circumstances organs are retrieved in a carefully orchestrated sequence. First the kidneys, then the liver and later the lungs. Once the other organs are taken it is no longer necessary to supply the body with oxygenated blood, so the heart too can be removed. But in the years preceding this first attempt at a face transplant in the United States, Pomahac had agreed with the New England transplant co-ordinators that the retrieval of the face should happen first, despite the fact that it wasn't a life-saving organ.

After removal, an organ can be deprived of its blood supply for only a short time before it too fails and dies. Measures are taken to extend that period for as long as possible: ice boxes and preservative solutions among them. But even with these the time that organs can survive without being plumbed into a recipient's new blood supply is limited to a few hours. And so the timing of the retrieval of Dallas Wiens' new face was critical.

As other transplant teams from around the region geared up to perform their retrievals, Pomahac received a phone call. A patient in urgent need of a heart transplant had been identified and there was no time to waste. The co-ordinators were clear: life-saving transplants took priority. Pomahac was told to leave Boston within the hour and that, upon arrival, he would be racing the clock. As he scrambled his medical team in Boston, his heart sank. Earlier in the year he had performed a partial face transplant; on that occasion the retrieval had taken six hours. Today he guessed he would have not much more than two.

This presented a huge challenge. The retrieval of a donated face is in many ways a task far more complex than the removal of the more familiar solid transplant organs. It must retain form and function and dozens of decisions must be made regarding what in the way of muscle, tissue and bone to take, and how.

Pomahac worked as fast as he could, with the other transplant teams beginning to circle. After two hours they had no choice but to ask Pomahac to step aside. With the retrieval of the face only partially complete, the removal of the solid organs had to begin. And when, finally, the heart transplant team left with their vital organ in hand, they took with them the blood supply to Pomahac's donor face.

From this moment onward the face was beginning to die. The team ran cold preservative solutions through its vessels to protect it, but this could only buy so much time. If they were not back at the Brigham with the new face connected to Dallas' circulation in less than four hours, all would be lost. Now the last surgeons left in the operating theatre, Pomahac's team worked furiously. When they finished, the face had been without a blood supply for over an hour and still had to be transferred by road and air back to the Brigham. It was going to be close.

~

Back in Boston, Edrich was making his preparations. The team was assembled, the theatre ready, its microscopes and surgical sets prepared. Dallas, accompanied by his grandfather, was ushered into the operating theatre suite. There was no time to lose; the face being brought by Pomahac would be fading. Dusky and starved of oxygen, its tissues were slowly dying. To survive it needed a new host and a new supply of blood.

Dallas was anaesthetised almost as soon as he arrived. Edrich sited drip lines in his veins through which to give fluids and drugs, and another line in his artery to monitor the blood pressure directly, with beat to beat precision.

By the time Pomahac reached the Brigham Hospital the skin of the donated face was a dull thundercloud grey – the colour of tissue and blood that has been stripped of its oxygen. It would not be long before it ceased to be viable at all.

Pomahac had to move quickly. He dissected out Dallas' external carotid artery. Having divided it, he pinched the free ends shut with an artery clamp. Then he began the delicate work of connecting that vessel to the face that he had just delivered. Working quickly, Pomahac threw stitch after stitch into place. And having made the connection he released his arterial clamp. For the first time in nearly four hours blood ran back into the oxygen-starved tissues. The face blushed pink.

∼

After twenty-one hours of paring back tissue, stemming dangerous haemorrhages and connecting blood vessels, muscle and bone, the operation was finally complete. The orchestra of surgeons withdrew and Edrich's anaesthetic team handed over to the intensive-care unit. But Pomahac, despite having been awake for nearly forty-eight hours, wasn't quite ready to retire. Once Dallas was settled on the ICU, Pomahac visited his side-room. After checking his patient's new face, Pomahac told the nurse that he was going to take a shower and change his clothes, but that he would be back to spend the night at Dallas' bedside.

'Dr Pomahac,' she said, with a smile, 'I think we can take it from here.'

∼

Dallas' new face wasn't immediately perfect in appearance. The tissues were swollen and bulky and the lines of surgical incision were evident. The face itself remained largely inanimate and insensate. Pomahac had expected all of this. It would take time before the full benefits of this procedure would make themselves known. But even in those early days it was clear that Dallas had been transformed. He bore almost no facial resemblance to the man who had been injured in that cherry picker more than a year earlier. But now, where there had been a blank canvas of skin, there were individual features: a nose, eyes, a mouth, lips and the more definite bony contours that make a face recognisable as being such.

Further sculpting of his features was necessary. Once the swelling had subsided, Pomahac trimmed excess tissue. Nerves and muscles needed time to become re-educated. But when Dallas returned many months later it was clear to all that the surgery had been a great success. His appearance was improved to the point where he might enter a room without anyone giving him a second look.

More impressively still, the nerves had begun to establish themselves. Dallas could now begin to express himself once more – re-learning how to smile and frown. He even regained his sense of smell. But most importantly of all, he gained sensation in the skin of his new face. For the first time since his accident he could feel his daughter's kisses on his cheeks.

~

The tale of Dallas Wiens' face tells us much about medical science's most spectacular triumphs, but it is in burns units all over the world that the everyday battles against fire are being won by slow and painful increments. Without McIndoe and his Guinea Pigs, the man lying in front of me now, body ravaged by burns, would have no hope of being restored to something of his former life. As it is, at least we can give him a fighting chance – providing we can get to the specialist unit in time.

We finally close the doors on our helicopter. The rotor blades spin up, the motors whining as they get up to speed. We rise backwards from the helipad with the idea that, should the overloaded engines fail, we have an outside chance of crashing into the helipad rather than the streets below. The oxygen levels in my patient's bloodstream continue to fall.

An aircraft alarm begins to ping in the cockpit. It is continuous and sounds malignant. I look around at the patient buried in wires and tubes, at the equipment we have jammed into the rear of the vehicle. The cabin is packed. We have wedged the gear in around the crew, myself and Louise the nurse. I imagine the strain on the engine and rotor blades. The ping of the alarm persists. It is, at this early point in the flight, a mystery to me how we'll manage to get our helicopter safely to its destination.

'Don't worry,' says the pilot as if reading my mind. 'Everything gets easier once you start moving forwards.'

CHAPTER 3
HEART

24 March 1917: the 9th Battalion, the Cameronians (Scottish Rifles) go 'over the top' during a daylight raid near Arras, northern France

THE WARD ROUNDS start at 6.30 a.m. That's the good news. The bad news is that the pre-round, when the intern visits all of the patients to prepare for the ward round, starts at 6 a.m. And I am the intern – here with one of the trauma surgery teams, at the very bottom of the hospital hierarchy.

It's the late 90s. The local drugs wars are coming to an end, but there are still plenty of guns around in Washington DC. The city jostles for position as the murder capital of the United States. Not so long ago there was an average of one murder a night in the square mile around the White House. This is not how I imagined the nation's capital.

I work twelve to fourteen hours a day. Every third day our team is on call. Those days I work from 6 a.m., through the night and the following day, thirty-six hours at a time. I have never worked so hard. I stay in a flat about half a mile from the hospital. I'd imagined that I'd spend my free time getting to know the city but I'm completely spent by the time the days are over. On the nights that I make it back to the flat, I force myself to stay awake long enough to make a microwave meal and eat it standing at the counter.

As time goes by, I become a little nervous about my neighbourhood. The next time I see my resident, Luis, I mention the shady characters I've noticed hanging around.

'Have you seen any shooting yet?' he asks. I misunderstand his question, thinking he's referring to our case load in the hospital.

'Sure,' I say. 'We see victims of shooting through the trauma rooms every night.'

'No,' he says in a thick Colombian accent, 'have you seen, from the window of your house, someone take a gun out of their pocket and shoot somebody else?'

'No,' I tell him.

'Then you don't live in a bad area.'

~

The crash pager goes off. 'Trauma call. GSW. ETA 3 mins' flashes the message on the screen. I climb out of the top bunk, trying not to tread on the junior resident in the bed below. GSW is the unit's standard abbreviation for 'Gun Shot Wound'. I hurry along to the trauma room. I'm not sure what time it is or how long I've been asleep. We arrive at the same time as the patient. The paramedics spit out a string of jargon: 'Eighteen-year-old female, GSW chest, signs of life on scene, arrested in the chopper, no output.'

She is wearing a blue dress with polka dots. Her feet are bare – presumably the shoes are somewhere at the crime scene – but she looks as though she's been out on the town. A pretty African-American girl, her makeup and hair done carefully.

The crew, who have just arrived on the helicopter, continue

with cardiac resuscitation while moving her to the trolley in the trauma bay.

'OK,' says Manish, the senior resident, 'let's crack the chest.' There is no equivocation; there is no time. The surgery must happen here; surgery of the most drastic and invasive kind. They must open the chest, expose the heart and lungs, look for a source of injury – a reason why her heart has stopped – and fix it. It needs to happen now.

A pair of scissors makes light work of her party dress. Another trauma nurse is getting the surgical trays ready. One of the junior residents is covering her torso in iodine solution as a hurried surgical prep.

Manish is taciturn at the best of times. The pace and gravity of the case don't make him any more verbose.

'Knife,' he says, levelly.

The handle of a scalpel is placed in his hand. Manish runs its blade across the skin, making an incision a couple of inches long in the side of her chest, just below her left breast. He pushes forceps into the exposed muscle, separating the fibres and creating a tract. He repeats this exercise on the right side of her chest. If air has become trapped in the pleura, the lining that surrounds the lungs, then its accumulation might be enough to stop the heart from beating. This is what Manish hopes for: that these simple holes in the chest wall might be enough to release trapped air and resuscitate the arrested heart. But tonight there is no such luck. He must proceed.

Manish returns to the right side of the chest and runs the knife along the line of the fifth rib, extending the incision he's already made until it reaches the breastbone. He divides the muscle between the ribs and then introduces the rib spreader – a gothic looking piece of stainless steel kit with a ratchet system that separates a pair of blunt claws, pulling the chest apart, separating the fifth rib from the sixth, exposing the contents of the chest cavity beneath.

And in less time than it has taken you to read this description, I am looking at her stationary heart and lungs.

Manish works quickly, inspecting the pericardium. This sack-like structure surrounds the heart like a glove around a hand. If it becomes engorged with blood, it will compress the chambers and stop them from beating. This too is relatively easy to remedy, but today it's not the source of our problems. There is blood everywhere in the chest cavity. A suction tube gurgles away. Manish shells the heart from its protective pericardial sac. He inspects it, hoping that there might be a simple hole, amenable to quick repair. But this is not the case.

He moves further up into the chest and at last finds the injury. A bullet has torn the great vessels surrounding the heart; her blood has been pumped out into her chest. The emptied ventricle has struggled and then failed to beat. There is no easy fix to this. The team stop resuscitating.

Manish asks the flight paramedics how long she has been down – without a pulse. More than half an hour is the reply. He looks at the clock on the wall and calls out the time of death.

The team leave. Manish first and then the other residents. The most junior member of the team is left with the task of closing the chest. I am that person. A huge curved needle on the end of a wire is handed to me. I am left alone with a girl who perhaps an hour ago was at a party when a man with a gun sprayed rounds into the room. One passed into her chest and through her heart and its surrounding vessels, leaving a repair too complicated to address in the short time that Manish had.

In the heat of the moment, during the resuscitation, it is easy to be objective about things, to separate yourself from the horror of the event. As a lowly intern you have a job to do, even if that job is to watch and learn, starting the process of preparation that gets you ready for the time when it might be you wielding the knife and making the decisions.

Alone with her, it's harder. As a student, the world of medicine appears to be full of patients who are much older than you, who are enduring things that you don't need to worry about just yet. But she is younger than I am, maybe even a teenager. She belongs back at the party, not lying here on a trolley with her dress cut to pieces.

The surgery she's undergone, an emergency thoracotomy, is a technique honed for precisely this situation. In skilled hands, in the right circumstances, it gives a victim of otherwise lethal penetrating chest trauma a 10% chance of survival. The immediacy with which it is brought to bear is startling. It is in some respects a simple, albeit violent, intervention. And today it wasn't enough.

Nevertheless, watching that procedure executed so deftly and with such surety, it is tempting to think of cardiac surgery as though it were an ancient art honed over centuries. But understanding how to open a chest and confidently operate upon the heart is something the surgical fraternity took almost the entire history of medicine to learn.

∾

If you place the palm of your right hand flat in the middle of your chest, its heel lying in the centre of your breastbone and your fingers extended so that your middle digit points at your left nipple, you can gain a good impression of where the heart lies anatomically. And while the beat of its apex is best felt well to the left, where the tips of your fingers resting on your chest now lie, the bulk of its mass is surprisingly central. Neither does the heart lie flat in the cavity of the chest; it is instead slightly rotated, its right side more exposed towards the front of the chest, its left slightly hidden to the rear. The whole arrangement sits protected behind the breastbone and a formidable cage of ribs; an evolutionary nod to the heart's central importance – and vulnerability.

From the breastbone, the route to the heart is two and a half centimetres in a straight line, but that trivial distance took medicine the best part of 2,500 years to travel. The twentieth century would see centuries of dogma set aside and cardiac surgery advance in great leaps and bounds. These feats of exploration lay open the continent of the heart to science and medicine in the same way as Scott and Amundsen paved the way to the Antarctic interior.

∾

Our exploration of the world's extremes is, in essence, an exploration of ourselves and the limits of the human body. It is our physiology, and our inability to protect it effectively from the physicality of the outside world, that put the remote corners of the Earth beyond our grasp until well into the twentieth century.

That exploration also saw us turn to the frontiers of medicine, to explore the limits of physiology in health and disease. The same revolutions in science and technology that extended our explorations of the physical world helped to push back the frontiers of medicine and surgery.

There were, at the start of the twentieth century, many facets of human anatomy and physiology that stood largely un-probed – foremost among them the human heart. While nineteenth-century scientists had begun to map the organ's function and complexity, it remained a territory upon which medicine still feared to trespass. As late as the fifth decade of the twentieth century, as the Second World War raged, the heart was still a continent as dangerous and unknown in the eyes of surgeons as Antarctica was to explorers of the heroic age.

Physicians saw the heart as largely inviolate: a sacred and complex whole that must remain intact and unaltered; an organ with which surgeons could not and should not interfere. This dogma was as old as Aristotle's teachings and remained unchallenged until the very end of the nineteenth century. Medical textbooks warned against tampering with the heart. In his 1896 text *Surgery of the Chest*, esteemed surgeon Stephen Paget made his position clear: 'Surgery of the heart,' he famously declared, 'has probably reached the limits set by nature; no new methods and no new discovery can overcome the natural difficulties that attend a wound of the heart.'

Overcoming the received wisdom of the past, making that leap of surgical faith, was a feat that required the terrible but unique catalyst of war.

~

It is winter, 1917. Somewhere on the Western Front a British infantryman is marching forwards across the frozen earth of no-man's-land. There is a blizzard in the air and a biting wind sweeping across the battlefield. His clothes are no match for this weather, but the crack of gunfire presents a more immediate threat. From the German trenches there is the sound of chattering machine-guns; the firing positions are perhaps five hundred yards away. At that range, in this visibility, there is a faint hope that their hail can be avoided.

The German machine-gun crews fire hundreds of rounds a minute, pausing only to clear stoppages, improve accuracy and prevent their weapons from overheating. Each bullet travels the best part of a kilometre in under a second. It spins around its long axis in flight, held stable by the same law of physics that keeps a child's spinning top upright – making the machine-gun accurate over large distances. But it is the velocity of the round, and the kinetic energy carried with it, that makes the bullet so lethal.

The soldier advances, a rifle in his right hand. His left is raised in front of his face to shield his eyes from the heavy drifts of snow. And through that blizzard, a spinning machine-gun round finds him.

The bullet travels through his left arm, just above his elbow, slowed by muscle and flesh. It continues on, exiting the arm, piercing first his tunic pocket and then its contents – a notebook and a bundle of letters – before encountering the wall of his chest and finally the substance of his heart.

At the time of the First World War, gunshot wounds to the heart were almost invariably fatal and cardiac surgery was still looked upon dimly. Back in 1883, Christian Albert Theodor Billroth, one of the founding fathers of abdominal surgery, had these words for would-be pioneers: 'a surgeon who tries to suture a heart wound deserves to lose the esteem of his colleagues.'

Views such as these continued to hold sway well into the Great War. In 1916, Major George Grey Turner, a doctor in the service of the British Royal Army Medical Corps, addressed an audience of surgeons bound for military duty. Although he had plenty of advice on other topics, he had little to say on injuries to the chest. 'These', Grey Turner told his audience, 'are commonly thought to be beyond the scope of surgery, and to merit very little attention . . .'

And yet the following year, in 1917, Grey Turner would receive a casualty recently returned from the Western Front, eighteen days after he had been injured by a machine-gun bullet – our infantryman with the bullet holes in his arm and his correspondence.

The soldier was in surprisingly good health and was indeed embarrassed to have been forced to arrive at the hospital on a stretcher. Grey Turner examined him and found evidence of a bullet's entry at the front of his chest but, ominously, no exit wound.

At the time of the First World War, medical X-rays were themselves a novel invention but their value in locating bullets and shrapnel was rapidly recognised and the practice quickly adopted by military hospitals.

The X-ray tubes at the disposal of hospitals of the time were primitive: a cathode and anode fashioned from metals such as tungsten, separated in an evacuated glass flask and driven by the electricity from an oil-powered generator.

That radiation, passing with ease through soft tissue, attenuated by denser bone and metal fragments, falling finally upon a fluorescent plate, revealed a bullet in the region of the heart's left ventricle. Further inspection showed it to be pulsating synchronously with the heartbeat and this was accompanied by a disturbing whirling motion. The bullet appeared to be lodged in the wall of the soldier's heart, its waist apparently plugging the muscle of the left ventricle, its tip inside the ventricle itself, wriggling with the flow of blood.

Grey Turner considered his options. Should the bullet migrate further it could lead to an embolus, showering fragments of clotted blood or infected material into the soldier's circulation, blocking distant arteries with unpredictable consequences. Or perhaps its dislodgement might lead to rapid and fatal haemorrhage. And even if it were to remain stable in situ, the presence of the bullet would surely be a source of disastrous infection. In Grey Turner's eyes there was little option but to intervene surgically.

The operation proceeded under a primitive anaesthetic cocktail of alcohol, chloroform and ether. Grey Turner made his first incision in the skin on the left side of the soldier's chest, shaped like a letter C, the size of a man's palm. Through this he removed the soldier's sixth rib and then divided the three ribs above it, allowing him to open the chest wall outwards as though it were the rear cover of a book. He retracted the lungs, gently pulling them out of the way, and finally gained access to the injured heart.

There he carefully opened the pericardium, the fibrous tissue sack surrounding and protecting the heart, from which the heart itself could be finally delivered.

～

The beating heart does not simply expand and contract. To witness it in life is to understand surgeons' traditional reluctance to interfere. There is an element of torsion in the way that it moves – waves spreading across its muscle from base to apex. Even in health its cadence constantly changes, accelerating and slowing periodically, but with a clear, intrinsic and vital rhythm. It exhibits a physical dynamism like no other organ in the human body and is inescapably the engine of life.

There is anatomical complexity too: tributary veins merging to converge on the right side of the heart in the vena cava, the last great vessel of the returning circulation. And this point of entry, the atrium of the right side of the heart, is but the first of four chambers. The second, separated from it by a fibrous three-leafed valve, is the right ventricle: thicker walled than the atrium above and able to provide a volume of blood with enough energy to see it through the pulmonary artery, its pulmonary valve and out to circulate through the blood vessels of the lungs.

That outflow divides and fragments into a plethora of smaller and smaller vessels, until finally they degenerate to form the many millions of fine capillaries, vessels smaller in calibre than the red blood cells that must traverse them.

Here those red blood cells, the all-essential vehicles of oxygen delivery, must distort as they squeeze through the capillaries, snaking around the alveolar air sacs, grabbing oxygen and giving up carbon dioxide as they go.

And then there's the return journey. Capillaries become venules, venules become veins until they too merge to comprise the pulmonary vein, returning the circulating blood once more to the heart. Here a third chamber, the left atrium, receives blood from the pulmonary vein, freshly laden with oxygen from its voyage through the lungs.

The left atrium provides just enough impetus to push the blood through a valve shaped like a bishop's mitre – the so-called mitral valve – whereupon it enters a fourth and final chamber: the left ventricle. It is this structure, with its muscular wall, that must develop enough force to accelerate the blood of a single heartbeat out through the aorta, past the aortic valve, to circulate around the body.

Through this network of vessels, into these chambers, over these surfaces, blood must flow endlessly, never faltering, never forming eddies or clotting, from moment to moment across the entire span of a human life. And, in 1917, somewhere in this complex mass of mobile, twisting tissue lies a bullet that Major George Grey Turner must find.

～

Grey Turner begins his search. He finds a scar covering an entry wound in the wall of the left ventricle. There is no doubt that the bullet lies within. He carefully cups the heart in his hands, trying to feel for the missile. The heart, he notes, develops the hardness of a stone while contracting, making it impossible to feel anything within. In the cycle of each heartbeat he has less than a second when the muscle is relaxed enough for him to locate the bullet. But even this proves impossible; the bullet is too deeply seated. Instead, he punctures the heart carefully but repeatedly in the area around the scar, remarking upon how solid its substance feels and noting that these needle wounds bleed but stop quickly of their own accord. In his written record of the operation, Grey Turner narrates his exploration like a mountaineer describing a new route. These are discoveries; territories uncharted. But the bullet remains elusive.

Unwilling to give up, Grey Turner pares the rib stumps back to give himself more room. He rotates the heart to examine its posterior aspect; whereupon – to his horror – it stops. Grey

Turner massages the now flaccid heart, squeezing it in his hands, hoping in some way to resuscitate it back to life. He rotates it back to its proper position, continuing to squeeze it in his hands, and finally it begins to beat again. But the bullet is still nowhere to be found.

After an hour and a half of searching in vain, Grey Turner decides to fall back upon the most fundamental of the general principles in surgery: *primum non nocere* – first, do no harm. To any practitioner of medicine, knowing when to stop is at least as important as having the courage to proceed. Admitting defeat, he withdraws and closes the chest, leaving the bullet lodged in the officer's beating heart.

It seems Grey Turner's instincts to withdraw were correct. The projectile was left in place and the patient duly recovered from his surgery. In fact, the soldier in question recovered fully and was even sought out and found by Grey Turner, twenty-three years after that abortive operation, in 1940, alive and well. His only complaint was of occasional fatigue; but that, Grey Turner explained, the patient had attributed to his exertions in 'the current war'.

∾

Grey Turner was not the only surgeon of the Great War to attempt cardiac surgery. Elsewhere there were reports of surgeries to remove missiles from hearts – some successful – but these were few and far between and not enough to convince the wider surgical fraternity that the heart could be reliably inter-fered with. The received wisdom of the time was to stand. The heart was perceived as all but inoperable.

But the Second World War saw the further mechanisation of combat; the practice of war became still more efficient and the spectre of wounded hearts returned. Shell fragments and bullets found their way into chests in greater numbers and casualties with wounded hearts once more began to arrive at military hospitals.

~

In 1942, aged just thirty-three, Dwight Harken, a captain in the United States Army and aspirant thoracic surgeon, returned for the second time in his life to England. He had previously visited London as a civilian, and had worked alongside the renowned British surgeon Arthur Tudor Edwards. But in 1942 Harken returned as a man of military rank, assigned to a post in General Hawley's office in Grosvenor Square, tasked with assisting the US Army in organising and co-ordinating medical logistics.

Harken hailed from the small town of Osceola in Iowa. Graduating near the top of his class, he won the opportunity to attend Harvard Medical School. Harken remained at Harvard as a graduate, spending part of his surgical residency in Boston and later New York, before winning a bursary from the New York Academy of Medicine to develop his interests abroad in a location and speciality of his choosing. Ambitious, but not wish-ing to compete with the likes of Allen Whipple, Edward Delos

Churchill and Elliott Carr Cutler – titans of general surgery – Harken decided to take a gamble and specialise in the newly emerging field of thoracic surgery. This was a bold move in an era that valued the gifted generalist more highly than the narrowly skilled specialist. Despite all this, Harken chose to travel to England and take up a visiting fellow's post at the Royal Brompton Hospital with Tudor Edwards.

Tudor Edwards was one of the few, if not the only, thoracic surgeon in the world at the time of Harken's secondment. His case load was principally concerned with the treatment of tuberculosis. Assisting in theatre, Harken marvelled at Tudor Edward's skills as a technician, watching keenly as he carefully explored the contents of his patients' chests and pared back tuberculous tumours, liberating blood vessels and elements of the branching bronchial tree from their encasement.

And yet Harken couldn't help but wonder why, when confronted with the diseased heart, which was in his eyes a mechanical entity, Tudor Edwards and his colleagues remained reluctant to operate, despite the pioneering work of Grey Turner's generation. So when the outbreak of the Second World War interrupted his apprenticeship with Tudor Edwards, Harken returned to Boston to begin his own experimentation.

~

Bacterial endocarditis, an infection of the inner surfaces of the heart and its valves, was an almost invariably fatal affliction in

Harken's time. In the absence of antibiotic therapy, the bacterial infection would disintegrate the heart's internal structures. Worse still, the pumping action of the heart would seed infection and emboli throughout the body. It was this formidable enemy which Harken sought to combat. In theory, surgical removal of the focus of infection would arrest the process and give the afflicted patient the opportunity to survive. But at a time when the world remained reluctant to enter the cavities of the heart, Harken's hypothesis needed the support of hard evidence before he could attempt it in human patients.

He began by working with dogs, but first had to build a model of the disease itself. This he achieved by operating on canine hearts and introducing metal clips to the surface of their mitral valves. He found that this intrusion always led to infection and the onset of bacterial endocarditis. This served two purposes: it provided a model of the disease he sought to treat, allowing him to simulate bacterial endocarditis in dogs, but it also gave Harken confidence that the cavities of the living, beating mammalian heart could be breached and repaired without immediate fatality.

However, Harken's work was once again interrupted by the events of war. As part of the United States' war effort, he was returned to England in 1943 and posted at Grosvenor Square under the command of Brigadier General Paul Hawley, Chief Surgeon on the European Theatre of War. Here, anticipating a flood of war casualties with penetrating chest wounds, he and Tudor Edwards campaigned for the establishment of specialist thoracic units, an endeavour in which they were successful.

In the first half of 1944, presumably in preparation for the imminent Allied invasion of Europe, several specialist thoracic units were set up throughout England. In May 1944 Dwight Harken was released from his office post in Grosvenor Square and, to his delight, assigned director of the 15th Thoracic Center at the 160th US General Hospital in Cirencester – a thousand-bedded hospital facility complete with a nearby runway to receive repatriated casualties from the battlefront. For Harken, this was a happy release from the burden of his administrative role at Grosvenor Square and one that would see him return to the operating theatre and resume his passion for surgery.

~

The US Army hospital was built in the grounds of Stowell Park in Northleach; it amounted to little more than a cluster of corrugated steel Nissen huts housing patient wards and surgical teams.

The month of May 1944 failed to provide much in the way of casualties to occupy Harken and his team. He spent the time productively nevertheless, preparing and training his clinical staff in the new art of thoracic surgery.

They would not have long to wait to put theory into full practice; 6 June 1944 – D-Day – was suddenly upon them. The hospital received a tidal wave of casualties, delivered by air from the European Theatre, first from the invasion and then a later second surge after the Battle of the Bulge. Confronted by casualties arriving with missiles lodged in their hearts, Harken consulted George Grey Turner for guidance on whether or not to attempt their removal. Grey Turner gave Harken his blessing, stating that there were many good clinical reasons to remove such foreign bodies but that the neuroses that might result from a patient's knowledge that he 'harbours an unwelcome visitor in one of the citadels of his well-being' might give cause enough. The challenge that Harken had so meticulously prepared for had finally arrived.

∾

One of Harken's great skills lay in understanding that the technical ability of the surgeon had to be matched with an equally capable operating team. These surgeries, particularly those involving foreign bodies in the cavities of the heart, often demanded considerable intraoperative resuscitation. While Harken navigated his way through the anatomy, his anaesthetist would be responsible for actively resuscitating the patient: providing massive transfusions; balancing efficient pain relief against the hazards of bleeding out, hypothermia and shock.

For the anaesthetist in these cases it was like flying a plane on fire, hoping that it could be held in the air long enough for the surgeon to be able to douse the flames.

Rates of blood loss of up to a litre and a half a minute were recorded – a volume loss that could empty the patient's heart and blood vessels and precipitate cardiac arrest in a matter of seconds. And while the physiology of shock was poorly understood at the time, Harken's team knew the value of massive whole blood transfusion in keeping patients alive.

Blood was supplied in glass bottles. But keeping up with the torrential losses demanded far more than their gravity-driven dribble could provide. To overcome the challenge of delivering blood at speed through narrow tubes, the anaesthetist would inject air into the head space of the flasks, increasing the pressure within and thus the rate of flow. Occasionally in the heat of the moment they would overdo it and the jars would shatter under the additional pressure, scattering shards of bloody glass across the theatre.

Harken meanwhile would be focused upon navigating safe routes to, and through, the heart. He came to learn that the simple act of handling the heart was enough to provoke abnormal and potentially fatal disturbances of its rhythm. And, like Grey Turner, he came to recognise the peril in removing the heart from its proper position. Harken would also devise techniques for incising and entering the heart while exercising at least some control over the resultant haemorrhage. This he achieved by placing sutures on either side of his incisions, leaving a pair of long trailing threads at both edges. His assistant could then hold these taut, keeping control over the opening in the heart as though it were the mouth of a purse. In this way, Harken was able to access bullets and fragments of shrapnel practically wherever they lay.

In the ten fraught months that followed the Allied invasion of Europe, Harken removed no fewer than 134 missiles from the hearts of wounded soldiers. The pace was relentless and the workload exhausting; Harken and his team would often operate around the clock for days on end, sleeping only when the lull in casualties would allow, in improvised cots. The demand for thoracic surgery outstripped the supply of adequately qualified surgical teams. Harken would sometimes operate by day and then travel by night, with his scrub team, to lend his thoracic expertise to other hospitals. And while the accounts of these surgeries were frightening, filled with stories of massive blood loss and tense moments, among the patients upon whose hearts Harken operated there was – incredibly – not a single death.

The effect on Harken of his experiences in Stowell Park was transformative. He had arrived in England optimistic but unsure that cardiac surgery involving the internal structures of the heart might be acceptably performed in humans. He returned to the United States at the end of the war convinced of this fact. And this time the medical profession would sit up and take full notice. The documented evidence was unquestionable: the heart was open for conquest. Major Dwight Emary Harken's explorations had proved it so.

~

The Second World War had been bracketed by not one but two awards for advances in antibiotic therapy. In 1939 the Nobel Prize went to German pathologist and bacteriologist Gerhard Domagk for his work in developing commercially available sulphonamide antibiotics, though the Nazi regime would forbid him from accepting it. In 1948, Ernst Boris Chain, Howard Florey and Alexander Fleming received their Prize for the discovery of penicillin. These developments would shape the future of cardiac surgery as much as any surgical technique. Bacterial endocarditis, which had hitherto been considered an unstoppable disease with a near 100% mortality rate, found itself suddenly amenable to treatment by the injection of antibiotic drugs. It was no longer the undefeated foe that Harken had so hoped to slay with surgery. But Harken's wartime experience had taught him that the heart could be opened and the mechanisms within, as he saw them, altered and repaired. He would turn his attention instead to diseases of the mitral valve – at the time, wild territory where respectable surgeons were loath to venture.

The mitral valve, seen from below as it opens into the left ventricle, has the appearance of a gently smiling fish mouth mounted on a ring of tissue around the size of a fifty pence piece. The delicately engineered mechanism is designed to allow blood to flow in only one direction, from atrium to ventricle. Without its system of valves, the heart is merely a pump which is as likely to push blood backwards as it is to push it forwards.

The leaflets of the mitral valve are prone to damage from the childhood affliction of rheumatic fever. Here, something as simple as a throat infection can lead to widespread inflammation and trigger the immune system to attack the body's own tissues. The resulting damage is akin to friendly fire – your body's own defences, unable to distinguish foreign invader from 'self', wreak havoc, attacking the skin, joints, eyes and the heart.

When this happens the mitral valve can become narrowed and the area through which blood can flow reduced, a condition known as mitral stenosis. As a consequence, pressure builds up in the left atrium which is transmitted back to the fragile circulation of the lungs. There, exposed to this unusually high pressure, the tissue-thin capillaries can fracture, spilling blood and fluid into the air spaces of the alveoli, causing coughing and breathlessness and the expectoration of blood-stained sputum.

While rheumatic fever is a disease of childhood, its cardiac consequences are usually seen later in life as the narrowing of the mitral valve progresses. But the physiological changes of pregnancy, which include an increase in the volume of blood pumped out by the heart every minute, can unmask the diseased valve. In Harken's time it was not uncommon to see young women present during their first pregnancy with the symptoms of breathlessness associated with mitral stenosis and heart failure. This condition became Harken's new target for surgical intervention. However, he was not the only ambitious young man determined to conquer this new territory, and it proved not to be an endeavour for the fainthearted.

～

Harken's first forays into mitral valve surgery were fraught with complications and loss. Six of his first nine patients died either on the operating table or shortly thereafter. After the sixth fatality, Harken's confidence was badly shaken, and it was only the intervention of his friend and collaborator Dr Lawrence Brewster Ellis that prevented him throwing in the towel completely. To complicate matters, Harken had competition on both sides of the Atlantic from the likes of Charles Bailey in Philadelphia, Russell Brock (later Lord Brock of Wimbledon) working at Guy's Hospital in London and Horace Smithy in South Carolina.

Within a year of the end of the Second World War, techniques in cardiac surgery had begun to advance all across the world. This was more than simple coincidence. Advances in the field of anaesthesia, radiology, blood transfusion and antibiotic therapy had conspired with the catalyst of war to usher the age of cardiac surgery into existence.

The contribution of these advances is often understated, as though they were not entirely essential to the establishment of elective cardiac surgery. But history had not simply waited for a surgeon bold enough to break with convention, or one with sufficiently gifted hands. The annals of surgery are, after all, replete with such individuals. It had been waiting instead for a means by which medicine might protect the brittle physiology of those with diseased or injured hearts from the added insult of surgery.

And so anaesthesia, antibiotics and transfusion medicine were together a primitive system of life support, a cocoon in which to wrap the patient ready for the onslaught of the surgeon's knife. Prior to the introduction of more carefully calibrated anaesthetic vapourisers and safer anaesthetic agents, it was not unusual for patients to die as a direct result of the unpredictable effect of the anaesthetic gases. These mysterious drugs had widespread and often deleterious impacts upon the body. They would cause profound falls in blood pressure but leave the heart over-excitable and prone to arrhythmias; they could precipitate respiratory arrest, cause hepatitis by inflaming the liver and provoke seizures.

The rapid and massive transfusion of whole blood, which itself had to be managed by a skilled theatre team, replaced volumes lost in haemorrhage, staving off shock and preventing hypotension and eventual cardiac arrest. And, in the period immediately after the operation, antibiotic drugs would keep infection of those profound surgical wounds at bay.

With a more stable platform from which to launch surgical interventions, the possibility of routine cardiac surgery became apparent to many. In the United States, Harken was thrust into direct competition with Charles Bailey, a surgeon of the same age, based at the Episcopal Hospital in Philadelphia. In the same year that Harken attempted his first mitral valve procedures so too did Bailey. And he would endure the same horrific rate of attrition.

Bailey's first patient, a thirty-seven-year-old man, had been incapacitated for more than ten years with mitral valve disease. His left atrium was, as a result, thin walled and fragile, rupturing during the operation before Bailey got near the valve itself. The patient bled to death in seconds.

During his second attempt at the same procedure, this time as a measure of last resort in a twenty-nine-year-old woman profoundly disabled by her narrowed mitral valve, he was able to access and operate upon the valve. He probed the valve at first with a surgical instrument, but having failed to dilate it sufficiently in this way, he decided to use his finger to increase the size of the opening. The patient died two days later from heart failure.

In the wake of these fatalities, Dr George Geckeler, Chief of Cardiology at the Hahnemann Medical College, wrote to Bailey: 'It is my Christian duty not to permit you to perform any more such homicidal operations.' And Bailey's record of failure had not gone unnoticed by colleagues and students. In fact, they began to call him 'The Butcher'.

Bailey would wait fifteen months to make a third attempt, this time at Memorial Hospital in Wilmington, Delaware, on a thirty-nine-year-old man. Again the operation failed and massive haemorrhage in the post-operative period was the cause of death.

Bailey chose to perform these later operations in a series of separate community hospitals, partly because he worried that successive failures would not be tolerated by any single centre. His fourth fatality, in a thirty-two-year-old man at Philadelphia General Hospital, would occur on the morning of 10 June 1948. The patient's heart became irritable and arrested as Bailey handled it. Despite frantic attempts to massage it back to life, the patient died on the table. Sensing that a moratorium would shortly be called on his procedure if he could not demonstrate something in the way of success, Bailey had already booked a fifth operation for that same afternoon. Leaving the Philadelphia

General he drove across town to the Episcopal Hospital for what he must have suspected was his last chance to show that this procedure had value.

At the Episcopal Hospital, Claire Ward was waiting. She knew the outcomes of Charles Bailey's previous operations, as did her family physician, who had advised her not to volunteer herself for the surgery. Claire was a twenty-four-year-old housewife. In childhood she had suffered with rheumatic fever and, over the years, the progressive narrowing of her mitral valve had led to mounting pressure in her left atrium and the symptoms of heart failure. The resulting disease had left her so short of breath that she could no longer look after her young child. If what Charles Bailey had promised was true, if his belief that the breathlessness and disability that plagued her could be abolished by this operation were correct, then for her the enormous risks were worth it.

By the time Bailey arrived at the operating theatre of the Episcopal Hospital, Claire was already on the operating table and being prepared for anaesthesia; whether he discussed the earlier events of the day with her is unknown. The anaesthetic induction, which had proved perilous in previous cases, went smoothly and once the patient was stable the operation proceeded rapidly. The pericardium was opened, the heart exposed, and sutures placed in the wall of the left atrium. Bailey incised and opened the heart, using first his finger and then a surgical knife to free the fused leaflets of the mitral valve. Satisfied that these manipulations had achieved the required result, he withdrew and closed the heart. The operation had taken eighty minutes. And this time it worked. Claire Ward left the hospital a week later, much improved. A month later she no longer had to take digitalis, the cardiac medication upon which she had previously depended. Ward went on to have two more children and lived for another thirty-eight years.

~

This first successful closed heart operation on a mitral valve was carried out by Bailey just four days ahead of his competitor Dwight Harken. Aware of this, and not one to be outdone, Harken raced ahead with writing up his own case report, making use of his contacts on the editorial board of the *New England Journal of Medicine* to achieve the much-prized priority over publication.

The rivalry between Harken and Bailey became the stuff of lore in the surgical community and provided something in the way of light entertainment for other colleagues. Their interactions were nothing if not ferocious. One resident came to write that the two would criticise one another fiercely and openly at surgical conferences and that Harken would become 'peri-apoplectic'. They were perhaps, at core, too alike in character to be able to tolerate one another. They were men of ambition and confidence who fully recognised the opportunity that lay at hand. Neither followed the dictates of conventional wisdom. They were born in the same year, attempted their first mitral valve surgeries in the same year and died in the same year.

The history of this era of surgery, in which the art made rapid progress but during which there were many deaths among patients, makes for difficult reading. It is unthinkable that any new surgical technique being pioneered today would proceed if accompanied by the same horrific rate of mortality.

It is tempting to regard Bailey and Harken as being so consumed by ambition and competition that they lost sight of the human cost of their endeavours. But while the pair did indeed race and compete, it's important to understand the complexity of the cases they faced. Physicians of the time had little confidence that the benefits of cardiac surgery outweighed its risks. In general, patients with diseased hearts were referred to Bailey and Harken only as a last ditch option, when they were already so critically unwell that there was little else to lose. In that context the fact that their failing physiologies often crashed completely when faced with the joint challenges of experimental surgery and primitive anaesthesia is perhaps less surprising. And Bailey, it seems, was driven by something more than simple ambition.

As a twelve-year-old child, he witnessed his father dying in his mother's arms of heart failure, breathless and coughing up blood-stained sputum. In his eyes an operation that could spare someone that fate was worth all that he, and his patients, would suffer in its development. His Christian duty, as he saw it, was not to stop but to succeed.

It's worth remembering that these events were also a product of their time. The absence of formally organised committees to oversee medical ethics contributed to a more permissive, less well-scrutinised style of practice. But perhaps also the war, so fresh in the collective memory, altered the perception of acceptable risk. Arguably society was more willing to accept sacrifice in the face of a war, whether against military foe or disease. Whatever your perspective, if it hadn't been for the dogged determination of the likes of Bailey and Harken, the fate of Claire Ward – and that of the many thousands of patients who followed in her footsteps – would have been very different.

~

Where does the line lie between innovative new therapy and experimentation? It is blurred at best, and pushing against the frontiers of physiology and medicine presents the physician with difficult ethical issues. Here the fate of explorers of the physical world departs sharply from that of our physician pioneers. Explorers risk their own lives; doctors risk the lives of those under their care, making their endeavour easier and at the same time infinitely harder.

But it was in this way that the last great chasm in all of surgery was crossed: the yawning breach of two and a half centimetres from the wall of the chest to the heart. The continent of the heart was finally open for exploration.

CHAPTER 4
TRAUMA

18 May 1969: a US medevac helicopter sets down in a tiny clearing on Hill 937 ('Hamburger Hill') as a wounded American soldier is rushed aboard

AT THE BATTLE of Waterloo, Wellington noticed a French doctor in the midst of combat, attending casualties and moving them quickly by horse and cart from the battlefield to the rear. Upon discovering it was Dominique Jean Larrey – Chief Surgeon to Napoleon Bonaparte – he ordered his men not to fire in his direction and, according to legend, lifted his hat in salute.

In medicine, the importance of speed has long been recognised. Larrey, present on the battlefields of the late eighteenth century, witnessed heavy artillery units wheeling and then retreating rapidly from the advancing enemy while casualties were left behind unattended. Only after hostilities ceased were the wounded collected and transported to field hospitals, introducing significant delays before definitive surgery could begin. In an age of more powerful firearms and artillery, surgery often meant the amputation of more than one shattered limb.

For those soldiers whose treatment was delayed for longer than twenty-four hours, Larrey noted that death was far more likely. His solution was to embed agile horse-drawn carriages with front-line troops so that the injured could be quickly carried from the field of battle during combat. Coupled with a rudimentary system of triage, in which the casualties were prioritised according to the severity of their injury, this innovation transformed battlefield care. Men whose traumatic injuries would have previously proven fatal were able to survive largely because of the speed with which they were attended to. The system was first trialled in the closing decade of the eighteenth century and its virtues were instantly recognised. Larrey's 'flying ambulances' were soon adopted across the armies of France.

Of course ambulances wouldn't actually take to the air until well into the twentieth century. The first helicopter evacuations of casualties took place in the Second World War, and by the time the Vietnam War arrived the sight of casualties being ferried through the sky to Mobile Army Surgical Hospitals – better known as MASH units – had become iconic. Civilian hospitals in the United States started to adopt these military solutions and began to use air transport to respond to trauma scenes – road accidents, shootings, stabbings and their like. And with this a new goal arrived. For the best chance of survival in the face of traumatic injury it was understood that treatment should begin as soon as possible. The concept of the 'Golden Hour' was introduced: the idea that after injury no more than sixty minutes should elapse before a patient received definitive care. The message was clear: delays in the treatment of victims of trauma were no longer acceptable. By the closing decades of the twentieth century, the process that Larrey had started during the Napoleonic Wars had led to a revolution in civilian trauma care.

~

In June 1998, having survived my brief stint as a student in Washington DC, I was back in the UK sitting my final medical exams. Having passed the written papers, I now faced the dreaded *viva voce* – an infamously cruel battery of oral examinations which saw candidates seated alone in small rooms facing interrogation by a pair of examiners. Sometimes they did the good-cop, bad-cop thing. More often you faced bad cop and bored-hungry-and-sick-to-death-of-examining cop.

Five of us shared a flat that year, all taking the final exams. Hugh was the first to go; we watched him head out to face the examining surgeons. He returned a couple of hours later in a foul mood. Though usually optimistic, he felt the day hadn't gone entirely to plan. When questioned further he revealed that he had fumbled his way through most of the thirty minutes, looking under-confident and less than knowledgeable. This, we reassured him, was OK and quite normal. But he wasn't finished. When the bell went to signal the end of the grilling he'd scowled, risen from his chair and left the room briskly, pulling the door sharply closed behind him as he went. This, he imagined, would send a signal to his examiners that he hadn't done himself justice and was usually much more impressive. But instead of the crisp clunk of a closing door he heard a startled cry. It turns out his examiner had followed close on his heels to call the next candidate in. And so Hugh, in his hurried theatrical exit from the room, had managed to slam his examiner's head in the door.

~

In July, Hugh and I stood on the steps of the medical school wearing capes, mortarboards and smiles that were impossible to wipe from our faces. It was over; we had graduated. We threw our hats high into the air and the official photographer clicked away.

Our house of newly qualified doctors moved into the Middlesex Hospital eight weeks later: twenty-four fresh-faced juniors, wide-eyed and terrified. We had studied for more than five years, endlessly learning and reciting the vocabulary and grammar that underpinned the art of medicine. We had learnt the language well enough; there was no other way to pass the final exams. But starting work was a challenge on a different scale, like going to live forever in the country whose mother tongue you had only just learnt to speak.

When you start medical school you assume that at some point you will go from being a know-nothing medical student to a confident healthcare professional. You think of the years of study as a chrysalis: in go spotty teenagers and, miraculously, out come doctors. But in reality you finish feeling more caterpillar than butterfly and if truth be told something of the caterpillar always remains.

In the first few days we received instruction on everything

from how to wash your hands to the correct way to complete a death certificate. They crammed the pockets of our long white coats with handbooks and too much information to digest, and then shoved us out onto the wards. For all of our training we were, at the beginning at least, worse than useless. We were guided not only by our senior medical colleagues but by nurses, ward clerks and hospital porters – all of whom, at that stage, knew far more about our job than we did.

Most evenings those of us who weren't on call retired across the road to the Cambridge Arms. There were better pubs but, to the exhausted house officer, this one had the virtue of being closest. We would drink too much, share the stories of the day and laugh at each other's most recent misfortunes and breathtaking displays of ignorance.

We quickly discovered that most of the job wasn't about life and death. Too much of it revolved around the filling in of forms and organising the list of patients. As the team's most junior doctor, your job was to keep a faithful inventory of the patients in your care, guiding your senior team round from bed space to bed space so that they could bring their experience and knowledge to bear.

Still, there were rare, terrifying occasions when you were temporarily alone – usually in the middle of the night, standing at the bedside of a blue and breathless patient. Knowing immediately that they were sick and that you alone didn't have the skills to stop them from crashing, you picked up the phone and called for help. After the phone went down you had a few hundred seconds of responsibility: a brief opportunity to make a difference – or at least to hold the line and prevent things from getting worse – before the cavalry arrived. Eventually help always came. Within the walls of the hospital you were never truly on your own.

~

As a student I learnt that 'trauma' derives from the Greek word for 'wound'. In the world of medicine it refers to injuries sustained as a result of violence or accident. That pair, violence and accident, are as ancient as the species itself. And to the non-initiate the mechanism by which they compromise the function of the human body looks like it should be easy to grasp.

Trauma is, after all, not the result of a bacterium invisible to the naked eye or a virus that subverts biology at the molecular level. Nor is it like heart disease in which unseen plaques of cholesterol lurk in the fat-laden vessels of the coronary circulation; or cancer, in which some arcane malfunction in the script of our DNA leads to the unstoppable division of a cell and its eventual invasion of our vital organs.

And so in training, when you're first introduced to the speciality of trauma, it is like a breath of fresh air: a disease entity in which the link between cause and effect appears absurdly clear. In your imagination it is little more than the transfer of energy to a mechanism and the disruption of its vital structures or the rupturing of its fuel lines. Only, like everything else in medicine, under the magnifying glass it turns out to be brutally complex.

Still, your first foray into the speciality of trauma is deceptively straightforward. 'Keep the oxygen going in and out; keep
the blood going round and round' remains the mantra. The
familiar boy-scout, ABC priority list rears its head here. Fix the
Airway first, then Breathing and lastly Circulation.

There are subtle modifications. While clearing and opening
the airway you have to pay attention to the spinal column. The
bones of the spine – the vertebrae stacked carefully on top of
one another – form a hard but flexible hollow tube which
protects the bundles of nerve fibres running in the soft tissue of
the spinal cord. If this bony armour has been damaged then the
cord will be vulnerable to injury and waggling the neck around
is likely to damage those nerve fibres and sever their connections, leaving the patient paralysed in all four limbs. For the
traumatically injured, protection and immobilisation of the
spinal column high in the neck are deemed as essential as the
letter 'A' in the list of ABC priorities.

Then, having dealt with injuries threatening the Breathing
or the Circulation, you're taught that in trauma the alphabet
goes a little further than 'C'. There is a 'D' and 'E' to look
after too. 'D' stands for Disability, and is a way of making
you remember to look for signs of injury to the spinal cord
by making sure that the power in the muscles and the sensation in the body's extremities remains intact. The 'E' is for
exposure, and is an *aide mémoire* to make sure that you have
looked from head to toe for hidden injuries. Casualties lying
on stretchers have been known to bleed to death from small,
penetrating wounds or scalp lacerations hidden from frontal

view. This is why trauma teams unceremoniously shear expensive clothing from trauma victims and then roll them naked onto their side.

Confronted with the worst traumatic injuries, it's easy to get distracted and miss easily treatable but potentially fatal injuries. The ABCDE approach is a tightly honed protocol, designed to offer a systematic approach to trauma that stops casualties dying from precisely this oversight. Properly adhered to and executed, this system, dubbed the Advanced Trauma Life Support protocol, will get you through the worst five minutes of even the most horrific trauma case.

At least that's the idea.

~

In those first few months we became marginally more competent and much less scared of just about everything. We got used to being exhausted; the worst weeks still ran to over a hundred hours. We lived and worked in the same building. Our dorm rooms lined a corridor on the fourth floor. And though we tried to make them feel like home, with scattered posters and plants, they were little more than a place to sleep. At this stage in our careers we found ourselves attached to the hospital as if by an umbilical cord. On the nights that we weren't working we rarely retired further than the pub across the road.

Weekends, we'd make a break for freedom. On a Friday afternoon, if you weren't due on call, you got yourself out of the hospital as fast as you could. Those that remained, holding the pagers and covering the emergencies, looked on with envy as colleagues fled the building.

On Friday 30 April 1999 it was my turn to be on call with the surgical team. It was the start of a bank holiday weekend, which served only to accelerate the usual exodus. I stood at the rear entrance of the Accident and Emergency department dressed in hospital greens, watching the ambulances come and go.

The A&E was windowless and always looked the same; open twenty-four hours a day, every day of the year, it was constantly illuminated by flickering fluorescent tubes and the glow of the X-ray boxes. Inside, the whirling hands of the clock on the wall somehow didn't give you a proper sense of the passage of time. For that you had to get out and glimpse something of the outside world, to catch sight of the fading light or breaking dawn. That was why we gathered in the quieter moments at the rear entrance to the hospital, with its less than spectacular view of the car park asphalt and the subtle aroma of ambulance diesel.

It was a fine and unusually warm evening. By half past six all but the most essential staff were gone. The department was quiet and the casualty waiting-room nearly empty. People, drawn out by the good weather, were packed into pubs and bars across the city and well on their way to being drunk. The injuries – the usual catalogue of ankles turned on cobblestones, assaults and road accidents – generally followed later, after people were kicked out onto the streets. I turned back into the department.

On the wall next to the nurses' station was the 'Red Phone', an old Bakelite thing with a dial. It only took incoming calls and rang with the old-fashioned trilling of a real bell. It was there so that the ambulance service could call ahead and tell us if they were bringing in something bad, giving us time to assemble and prepare in the big resuscitation bays at the back.

Just before a quarter to seven, the Red Phone rang. Alex, the nurse in charge, listened intently, scribbled some details and put the handset down. Usually a pithy case summary about the imminent arrival of a single patient would follow: a heart attack, a massive drug overdose or perhaps a stabbing. The department was set up to tackle these blue-light emergencies without breaking its stride. A small team would peel off into the crash rooms and get on with the resuscitation while the ingrowing toenails, superglue accidents, coughs and colds continued to stream in through the front door.

But this call was different. Alex raised her voice to make herself heard. There had been a bomb, she told us, in a nearby pub in Soho. There were many casualties. A major incident had been declared so that we could prepare to receive casualties. That was all the information we were given.

We all stopped for a moment, trying to digest the news. Then the sound came, with heart-stopping strangeness, of every pager in the hospital bursting into song simultaneously. A machine-gun succession of monotone chimes was followed by the crackling of the pagers' tiny loudspeakers. 'Major Incident Declared,' came the slow and deliberate voice of the switchboard announcer, and then again: 'Major Incident Declared.'

The trauma team assembled in the crash rooms: a surgical

registrar, a senior house officer and me; our consultants were already on their way back to the hospital. The Red Phone rang again. There were many people injured and some trapped. The ambulance service was asking for a mobile team of doctors and nurses to go out to the scene.

The surgical registrar didn't want to lose surgeons who would likely be needed to run the resuscitation bays or assist in theatres. I, on the other hand, could be spared.

The mobile team consisted of three doctors and three nurses. The accident and emergency nursing staff, familiar with the major incident drill, manhandled us through the preparations. I was pushed into a kit room that I had never before noticed. I found myself pulling on a fluorescent suit and donning a hard hat. A trauma pack, full of equipment and drugs, was shoved into my hands. And then I was being ushered out into the ambulance parking bays. Before anything more could be said, a pair of ambulances screamed into the parking lot and flung their rear doors open. The six of us climbed inside.

I sat clutching my trauma pack next to an A&E registrar and opposite Christine, one of the nurses. The rest of the team was in the other ambulance. Christine leaned in and raised her voice above the noise of the sirens: 'This is a major incident,' she said. 'You are wearing your personal protection equipment. In your jacket pockets and pack are the following items . . .' She started reeling off the list.

I just stared at her. She was looking me right in the eye and talking to me – nearly shouting, in fact, to get above the din of the engines and sirens – but it felt like she was addressing someone else. She paused, waiting for me to carry out my checks. I pulled the pockets open. It was all there, just as she'd said it would be. It felt like a conjuror's trick. How could she know what was in my pockets? And then the realisation dawned that she, unlike me, was trained for this event. She, at least, knew what she was doing.

The ambulance went up and down through its gears, hurtling towards the scene with us in the back. The journey couldn't have taken more than five minutes. When it drew to a halt I didn't really know where we had arrived. And then the rear doors opened.

It took more than a few moments to orientate ourselves. We were at the junction of Old Compton Street and Dean Street, part of London's theatre district. The streets there usually thronged with tourists and clubbers. But as the doors opened it was unrecognisable. Glass carpeted the street, ambulances crowded all of the access roads; the place was awash with casualties.

When we arrived the scene was barely under control. The bomb had gone off less than thirty minutes earlier. The pub was still smoking; the fire brigade had just finished hosing bits of it down. A registrar from the Helicopter Emergency Medical Service (HEMS) had arrived ahead of us, and had taken charge of the strategic medical decision-making. He, along with the ambulance officers, had separated the walking wounded from those most severely injured.

We were taken to a litter of bodies in the middle of Old Compton Street being attended by paramedics and firemen. They were the most severely injured people I had ever seen, and remain so to this day. There were perhaps a dozen people on the ground in front of us. They were exposed; most of their clothing had been incinerated in the flash fire that followed the detonation. Their bodies had suffered burns; some of their limbs had fractured or been amputated by the blast.

In the worst situations the trick is not to think too hard. It's best to stay focused on the one task at a time and best to make that task as simple as possible. For fast-moving situations there are protocols that can be unpacked and delivered almost reflexively. And while these achieve many things, one of their most important functions is to stop you – struggling in the midst of events that words could never adequately describe – from grinding to a halt.

It's here that the alphabet of survival comes to the rescue, that system of Advanced Trauma Life Support. Those simple ABCs get you moving, stop you thinking. You don't much care about where the system came from or how it was designed. You are just grateful that it exists and you wonder how anyone could be expected to cope without it.

～

Back in 1976, Dr James K. Styner was flying his family from Los Angeles to his home town of Lincoln, Nebraska, after attending a wedding. The family had travelled some considerable distance: east across Southern California and Arizona, landing briefly in New Mexico to refuel.

They took off and continued their journey home, eventually turning north over Texas and up through Oklahoma and Kansas. As they crossed over into the airspace of their home state, Nebraska, they ran into thin, low-lying cloud. The day was coming to an end and, not being rated to fly on instruments alone, Styner chose to stay below the cloud base. By the time they reached Lincoln, the Sun had already set but they were almost home. Then – flying low with the skies nearly dark – Dr Styner became suddenly disorientated. He steered his plane low across a pond and realised too late that he was below the treetops.

There was a deafening roar as the aircraft hit the trees and ploughed into and then across the ground, ripping through the underbrush, disintegrating as it went. In those brief seconds James Styner waited for his life to end. The wings were torn off almost immediately and the remaining fuselage slid for over two hundred feet, spinning so that it finally faced backwards. The plane somehow came to rest upright, its fuel tanks ruptured and spilt, a huge hole gouged in its right-hand side.

James opened his eyes, amazed that he had survived the impact. The lower ribs on the left side of his chest were fractured and his forehead and face, which had smashed into the dashboard, were deeply cut. Charlene, his wife, was nowhere to be seen.

The world fell silent around him. Dazed, he pulled himself out of the aircraft. Once outside his head cleared a little and his priority became the children still trapped in the wreckage.

Chris, ten years old and the eldest of the four, was least injured – his arm was broken and his hand was bleeding but he was still awake, alert and orientated. Kim, sitting on his lap with the same belt around her waist, was unconscious, her head having collided with a fire extinguisher. She was three years old. The other children, Rick and Randy, were in even worse shape. Eight-year-old Rick's head was deeply lacerated and he too was unconscious. And Randy's leg was partly impaled on the jagged fuselage and trapped beneath the plane.

Dr Styner got Kim and Rick out first. Then he returned to the plane and dug into the ground around Randy's trapped leg; excavating and freeing it from its impalement. He expected it to bleed profusely, but mercifully it didn't. Chris, with his broken arm, managed to find his own way out.

They gathered some clothes from the scattered luggage and piled them like blankets over the younger children. It was winter in Nebraska and that night temperatures fell below freezing. They waited in the darkness for help that never came. Finally, realising that they were on their own, James went out in search of his wife Charlene. He ventured out twice without success, returning to the children each time. On the third occasion he found her. She had been thrown more than three hundred feet from the plane and had suffered a catastrophic head injury. Charlene was dead. With temperatures still dropping outside James now had to focus his attention on his children.

Worried about the injuries they had suffered, Dr Styner decided to go in search of help. From the crash site they could see a road in the distance. He was aware of pain in the ribs over-lying his spleen and wondered if it too might be injured and bleeding or in danger of rupture. If that were true then the long march in search of help would only make things worse. But with no idea when, or indeed if, rescuers would arrive, he decided that he should go anyway.

James talked frankly with his ten-year-old son Chris. He told the boy that he was concerned about the state of his own spleen but more worried about his children's injuries. He explained his plans to go and get help and said that if he didn't return, Chris shouldn't go looking for him but instead stay with his brothers and sister. His voice was calm and remarkably free of emotion. James said goodbye and then shortly after 2 a.m. he set out for the road.

After what seemed an eternity, he finally reached the road and flagged down a car. His face was caked with blood and initially the occupants were hesitant to leave their vehicle. But he managed to explain his situation and together they returned to the crash site. They gathered up the children and Styner said a final goodbye to his wife. Then, somehow, the five of them crammed into the back of the car and drove a few miles south to Hebron hospital.

Hebron was a small community hospital and when they arrived in the early hours of the morning the door to its Emergency Room was locked. A lone night nurse stood at the door and asked them to wait for the doctors to arrive. Somehow they forced their way in, but things didn't improve much when the hospital's medical team finally arrived. Their approach lacked structure and seemed to ignore key injuries. It became obvious to James that they were unprepared for the nature and extent of the family's injuries. Sliding off his trolley, he stopped the local doctors from treating his children any further and took over their care. He had come too far for things to fail here.

Next he contacted colleagues in Lincoln and organised transport, by air, back to his own hospital. They landed at Lincoln Airport and travelled by road to Lincoln General Hospital's Emergency Room, arriving at 8 a.m. – more than fourteen hours after their crash. There James Styner could finally resign the role of doctor to a team of his friends and colleagues and once more become a patient and parent.

Dr Styner was incensed by how long it had taken to get his children the trauma care they needed. He didn't blame the physicians and nurses at Hebron, but he felt that he'd been able to deliver better care as a trauma victim at the scene of the accident than he'd received at the local hospital. And if that was the case, the system was broken and things would have to change.

In the years that followed the accident, James K. Styner invested all of his efforts in designing a straightforward protocol for the management of trauma cases, one that could, if necessary, be delivered by even the smallest of hospitals. He based it on existing models for the delivery of cardiac resuscitation, adopting that powerful ABC approach and extending it. Just four years after his plane crash, Dr James K. Styner's Advanced Trauma Life Support (ATLS) course was adopted by the American College of Surgeons. He trained people to deliver life-saving trauma care whatever their situation and then he trained them to train others. Courses sprang up all over the United States and then all over the world. In the years that followed, ATLS went viral. To date more than a million people have learnt to follow it. In a 2006 lecture, Styner told the remarkable story of its origins and finished by joking that it had spread around the world and would soon be taught on the Moon and Mars. He wasn't far wrong.

❧

Right there on the chaotic Soho street, I checked the first of the casualties I came to. He was lying on the ground, his clothes in tatters, his skin scorched. There were nails embedded in the skin of his chest and abdomen, but his hands were warm and he could still talk to me clearly. I pushed an intravenous drip into a vein on his arm. I tried to stick a dressing over the top of it but it wouldn't hold; the layers of burnt skin just sloughed off underneath. I'd never seen anything like it. Grabbing a crepe bandage from my pack, I wound it around the line and tied it in place. And then, having done the little I could, I realised I had to leave him with the paramedics and move on to someone else. I turned to discover a much more seriously injured man with Christine already at his side.

There was bleeding; at least that's what I remember most of all. One of his legs was missing and his face and chest were burnt. Bits of shrapnel protruded from his remaining limbs. He was awake but only just. I started at the top, at A, checking his mouth for injury or obstruction. Then on to B: I got my stethoscope out and went through the motions of placing its bell on his chest. But above the chaos of the scene neither the gentle rush of air nor the drum of his beating heart were audible. I put my cheek close to his mouth and turned my head to look at his chest, watching for its rhythmic rise and fall and the rush of warm air against my face. At this point one of the firemen pointed out the amputated leg; worried perhaps that I was ignoring the obvious injury.

He was right of course. When it comes to trauma the alphabet arguably should start at C for Circulation Major haemorrhage has to be dealt with first. The heart circulates around five litres of blood a minute, more if you've just been injured and there's a lot of adrenaline around. For a man of average height and build the whole circulation holds perhaps only five litres and so a significant bleed will kill in minutes – at least as fast as an obstructed airway or injured chest. I looked at the leg. There was plenty of blood on the floor and it appeared to be oozing steadily. I felt down for the femoral artery in the front of his right thigh. The pulse there was still good and reasonably strong. I grabbed one of the firemen and pulled his hand down onto the spot where I could feel the pulse, asking him to push down hard over the artery with his thumb, hoping that this would close it off and slow the rate of blood loss. Then I carried on with my survey, working again from head to toe.

I sited a line into his arm too and started some fluids. He was sick and getting sicker. There in the middle of the street there was nothing else I could offer, nothing else I knew how to do. And just at the point at which the protocol I was following ended and I might have started to flounder, one of the HEMS paramedics walking by put a hand on my shoulder and said: 'Does he need to stay or go?'

'He needs to go,' I said.

~

It is tempting to think of that as the end of the story for trauma: the point at which the ambulance doors shut and the victim is sped along to the nearest hospital. And in truth, rapid access to treatment and the ATLS protocol have transformed the survival rate of seriously injured casualties. But the fight doesn't finish there.

For Dominique Jean Larrey on the battlefields of the Napoleonic Wars, trauma surgery relied largely upon getting casualties into a field hospital as fast as possible and then stitching wounds and performing amputations promptly to arrest bleeding. There were no anaesthetics or antibiotics, and survival depended upon addressing the primary injury before it became truly life-threatening.

Today, for the patients who arrive in our trauma units, we can do a lot more: we fill them with blood, splint limbs, throw stitches in the ruptured vessels we can see, scoop out bleeding spleens, repair punctured viscera and pack lacerated livers. All this we do while holding their physiology stable with drugs and life-support machines.

And innovation in the pursuit of survival has taken us further still. We use beautifully nuanced physics to image, in fascinating detail, what remains hidden. We can thread long tubes into vessels, snaking them up from distant points of entry. These we use to deploy devices that block or stent torrentially bleeding arteries and veins that can't be reached safely or quickly with a knife.

But once the patient is taken from the scene of the accident and resuscitated, once they've been opened up and their haemorrhage stopped, the fight continues. It's not enough to understand the mechanisms of bleeding vessels, crushed viscera and fractured limbs. Having turned off the bleeding taps, having saved limbs and organs, patients can still continue to decline.

For the most severely injured patients, kidneys can go on to shut down, hearts can malfunction and lungs can fail. It is this secondary wave of illness that follows in the wake of major injury that takes trauma out of the realm of the simple, physical disruption of a mechanism and turns it into a complex and formidable disease.

~

Why do you bleed to death? That's the sort of seemingly innoc-uous question they dish out at finals. And you fall for it hook, line and sinker, relaxing because the answer must be blindingly obvious. But it's like someone asking you how a wheel works. You think about it for about ten seconds and then realise two things: you don't know and you never did know. If there's no blood running in your veins, you die. Yes; but death, or at least cardiac arrest, happens long before the system is empty of blood. And so the question becomes what causes the cardiac arrest? And suddenly that innocent question starts to be about what determines the force and energy of every heartbeat. And then you know you're in trouble.

Nerves running from the brain send impulses to the heart, moderating its pace and force of contraction, and at times of stress adrenaline circulates in the bloodstream driving it harder and faster. This is what next comes to the mind of the desperate candidate. But in the face of injury and massive blood loss, it's not the malfunction of this system that causes the heart to falter and stop. It is the mechanism of the heart itself. The heart has specialised muscle fibres which match the force of their contraction to the amount of blood entering it. If more blood returns to the heart just before the start of a heartbeat, the muscle contracts harder and so pushes out a larger volume. It's a way of making sure that the heart ejects the same volume of blood as enters it. If it were unable to match its ins and outs in this way, it would rapidly balloon and fail.

But when less blood returns to the heart it beats with less force. And if the circulation is suddenly losing volume because of a haemorrhage, the heart empties further and its contractions become weaker. If bleeding continues unchecked, the heart eventually arrests.

In the first moments of treating a trauma victim, it's precisely this that you're trying to prevent. It's why it's so important to stop the bleeding and maintain an adequate circulating volume in the blood vessels. And after the first phase of resuscitation is complete, after you've followed James K. Styner's all-important ATLS alphabet to the letter, much of your effort continues to be dedicated to the same task: stopping hidden bleeding with surgery and restoring volume with blood and fluids.

But even without resuscitation, the body starts to protect itself. And it uses the same strategy. In the face of massive haemorrhage, injured blood vessels spasm and shut themselves off to prevent further loss. Elsewhere vessels in the extremities constrict, forcing blood back towards the central, vital organs, temporarily depriving less important tissues but returning more blood to the heart. This reflexive recoil of peripheral capillaries near the surface of the skin is partly what accounts for the pale appearance of trauma victims. But the body's response to trauma goes well beyond the heart and its system of blood vessels.

Hormones pour into the bloodstream, mobilising fuel stores from fat, raising the body's sugar levels. The protein in muscle too begins to be broken down and its constituent components recycled to assist in the defence, like a country putting its economy onto a war footing.

But the most complex and problematic aspect of the human body's response to major trauma is that of the immune system. At the site of injury, white blood cells patrolling in the bloodstream and the cells lining injured vessels and tissues release messenger molecules. These summon a host of other immune cells which take part in removing dead and damaged material and prepare the way for healing. In moderate injury it is a beautifully orchestrated process which sweeps away cells that are no longer viable and replaces them with new ones, all the while making sure that enough energy is made available to cope with the increased metabolic demand of this restoration.

These are mechanisms evolved over millions of years. They are what allowed our ancestors to hunt and to defend their families – to be predator and prey – and then limp off into the bushes and return rebuilt.

And so for minor and moderate injury – deep lacerations and wounds that don't involve vital organs or uncontrolled blood loss – the body's response has been carefully honed over Darwinian timescales to work in our favour, assisting our survival and ensuring that our genes continue.

But for severe injuries, of the type that would have killed a person quickly in the days before modern medical intervention, no appropriate survival process has evolved. Instead the immune response to trauma oscillates wildly, causing more harm than good.

The immune response is effectively a very powerful and potentially very destructive chain reaction. It's much like firing up a nuclear power station. You want to encourage the reaction to kick off, go critical and generate heat, but you have to moderate it well enough to avoid meltdown.

And that's exactly what happens in the immune system. Under-activation would leave the victim prone to infection. Over-activation would lead to the malfunction of our organ systems: the biological equivalent of meltdown.

There is no evolutionary precedent for the limits of survival we are now probing. By the time we're supporting multiple organ systems on an intensive-care unit in the wake of major trauma, we've left evolution far behind. Out at those extremes we depend not on our physiology, but upon state-of-the-art systems of life support and the speed with which they can be brought to bear. The idea that, in the event of major accident, a team might literally drop out of the sky, scoop you up from the road and propel you within minutes to a hospital, is a construct of modern medicine that has only existed in recent decades. The edge of life, in that respect, has never been more heavily invested in. And expectations of survival in the face of horrific physical injury and physiological insult have never been so high. All of this means that today, when faced with even the most extreme trauma, we are less willing to accept defeat.

∿

When the bombsite was clear of all the major casualties, we moved up to Soho Square to check over the walking wounded. We walked along Dean Street. The cafés and bars were entirely vacant, their tables covered with half-eaten meals and hurriedly left drinks. Afterwards we returned to the hospital and I worked through the night.

The pub that had been bombed was called the Admiral Duncan. It was a bar popular with gay men. The bomber, a twenty-three-year-old paranoid schizophrenic and former member of the British National Party, had planted and detonated two previous devices in the preceding fortnight – one in

Brick Lane and one in Brixton. The bomb in Soho was the first one to inflict fatalities.

~

On Friday 30 April 1999 Andrea Dykes and her husband Julian travelled to London. John Light, who had been best man at their wedding, was with them. Andrea was newly pregnant and they were in celebratory mood; John was to be godfather. They were on their way to the theatre to watch the musical *Mamma Mia* but decided to stop for a drink at the Admiral Duncan. There they were joined by John's friend and former partner, Nik Leer.

The bomb detonated at 6.37 p.m. More than a hundred other people were injured. The blast killed Andrea and Nik instantly. John died the following day. The seven other casualties admitted to intensive care all ran difficult and prolonged courses, but all of them survived.

On 30 June 2000, David Copeland was found guilty of the murder of Nik Leer, Andrea Dykes and John Light. In 2007 the High Court ruled that he must serve at least fifty years in prison.

CHAPTER 5
INTENSIVE CARE

April 2003: Passengers on Hong Kong's Mass Transit Railway (MTR) during the Severe Acute Respiratory Syndrome (SARS) outbreak

THE LINES ON the electrocardiogram (ECG) stumble again, the complexes on the screen tripping over one another, degenerating once more from the essential electrical rhythm of life, becoming something more lethal. He is eighteen years old and I don't know what's wrong. The nurses charge up the defibrillator again. We deliver the shock. There is a pause. The cells of the heart reset themselves and then a more normal rhythm returns.

We have taken blood, shot X-rays and run CT scans in search of a diagnosis. There is little there to guide us. We have examined him from head to toe. His chest is clear, he is free from injury and his kidneys appear to be working – at least for now. But his blood chemistry is a mess. Organic acid is building fast – faster than his kidneys and lungs can clear it. Lying there in the bed, unconscious and ventilated, surrounded by the blinking lights of enough monitors to put a Christmas tree to shame, his heart driven by drugs, his lungs driven by a machine, he has the physiology of a man exhausted and on the verge of death. The ECG degenerates once more. We shock again.

His belly is slightly swollen. Perhaps there is a problem with his gut. Perhaps, somehow, a branch of the circulation that supplies the loops of bowel has become obstructed or compromised. That would be more than enough to make him critically unwell. But he's really too young for that to be likely. We review the CT scan images. To our eyes they are unremarkable. None of it adds up. We call in the surgeons. They are reluctant to operate. If they take him to theatre, he'll probably die on the table. But if we do nothing he will die for sure. We debate the decision and while we do so I shock him again. It is perhaps the tenth defibrillation. I have lost count.

This is intensive care. We can support hearts, replace kidneys, ventilate lungs. We can resuscitate, render unconscious and replenish. This is the sharp edge of all that can be done to support human physiology against illness and injury. This is everything we have, and still I cannot see how we can possibly win. When is enough enough? Perhaps the surgeons are right. It is after all absurd – extruding a man's physiology to its very limits in this way, well beyond any realistic expectation of survival. Why should we set ourselves against these catastrophes, when there are other fights that might be more easily won?

The formidable systems of artificial life support at modern medicine's disposal create new problems. The desire to find something more that we could do in the struggle to save life is sometimes replaced by the need to understand when to stop. To help understand why we try at all – and the events that gave birth to the first intensive care units – we must first go back to a time and place where modern medical interventions would have seemed like the stuff of science fiction, and technology presented little obstacle to death.

~

The village of Grand Gaube, on the tiny island of Mauritius, is set back inland only a few hundred feet from the Indian Ocean. In 1946 it was a ramshackle collection of the most basic dwellings, separated from the sea by a beach of brilliant white sand. My father, Ah Yoong, and his family lived in a single-roomed hut. He was nine years old and shared the floor space with his parents, his two brothers, Daniel and John, and two sisters, Angele and Therese. The roof was made of corrugated iron and the walls were made of stone with barred openings that served as windows. It was, in my father's estimation, the best house in the village by far.

His parents, Li Moon Ki and Tan Tin Ying, were immigrants from China, finding their way from the South-Eastern Chinese province of Guangdong, via the ocean-going trade routes, to Mauritius. They were Hakka people – literally 'the visitors' – nomadic over centuries, moving where the land was good, never limited by geographical boundaries. When the age of steam came they boarded ships in search of prosperity. That journey ended in Mauritius, a tiny volcanic island maybe thirty miles wide and not much more in length, fringed by white beaches and a vibrant coral reef. Tan Tin Ying was by all accounts a woman of fierce character and intelligence, but she was illiterate. Li Moon Ki, however, was among the few men of the village who could read and write. The house doubled as a general store, selling everything from rice and spices to liquor and nails.

Grand Gaube was a fishing village, a ramshackle assembly of huts with wooden walls, thatched roofs and cow dung floors.

There were outside standpipes bringing fresh water, but only the most basic sanitation.

For the residents of Grand Gaube, the sea was their life. They took its spoils and were hostage to its temperament. And they were vulnerable to the tropical storms it brought, particularly its cyclones. In the summer of 1945, two cyclones passed near by Mauritius and a third descended on the island itself. These spiralling winds, with gusts of over a hundred miles an hour, carried drenching rains and destroyed what little infrastructure villages like Grand Gaube had. Afterwards my father and his siblings collected the fish freshly strewn along the beach, and swam in newly formed pools, brought by the storm and tides that had run inland. But sewage had spilt into these waters, and disease swiftly followed.

That summer an epidemic of polio broke out across Mauritius, causing as much devastation as the cyclones themselves. The virus causing the disease could be carried in the gut and then spread in faeces. Poor hygiene, the destruction of infrastructure and bouts of diarrhoeal illnesses following the cyclone all conspired to amplify the spread. A team of British epidemiologists tracked it as it moved from village to village, often carried by healthy adults who'd built up an immunity to the virus.

What followed is an example of what happens when a transmissible, disabling and potentially fatal disease encounters a population with only the most rudimentary public health provision. During that summer there were more than a thousand cases of poliomyelitis on the island. The children were by far the worst affected. Of 851 cases identified and recorded by epidemiologists, around two-thirds were under the age of five and more than 90% were under ten. The virus was aggressive and unfettered by modern medicine. Almost every case identified by the epidemiology teams – nineteen out of every twenty – suffered paralysis and withering of one or more limbs.

In my father's family, his older sister Angele was the first to fall sick. For days she suffered with high fevers and drenching sweats. Grand Gaube had no doctor of its own. Occasionally a physician would pass through the village but he was seen as a charlatan and viewed with distrust by most of its residents. And so Ah Yoong was sent out by his father to pick the leaves of the lilac tree, from which a cool bed could be made, insulating Angele from the hot floor in the hope that this would somehow reduce the fever. But the fever continued and Angele appeared to be getting weaker.

In the earliest days of the illness, Ah Yoong would take his older sister by the arm to help her walk. Later he resorted to carrying her on his back.

Eventually the fever passed but Angele was left paralysed, unable to walk. She was just nine years old.

~

The basis for the conscious process that triggers us to move a limb, speak a word or register a thought remains elusive, and likely will for some time to come. Consciousness is the last dark continent of life science. We are incapable of properly defining it, much less understanding how it works.

But the processes it sets in motion are better understood. When it comes to movement we know that the motor cortex is the point of origin of signals that trigger voluntary movement.

You can get an idea of the location of this thin strip of brain by putting your thumb on your earlobe and then stretching your index finger up until it reaches the top of your skull. Below the quarter arc now made by your finger and thumb, beneath the layers of skin, bone and tough protective tissues, lies a narrow strip of brain. It is less than a centimetre wide and penetrates to only a few millimetres below the brain's surface. And yet in this modest layer lies a population of pyramid-shaped cells from which the impulses that initiate movement are first fired. Those nerve cells are neurons, responsible for connecting thought to action, specialised for the task of transmitting signals from brain to muscle bed.

Most of us could have a good crack at drawing an animal cell. You'd start with an indefinite oval and somewhere near its centre you'd plant a circle which you would shade in and call the nucleus. A couple of smaller scribbles around that nucleus would give you mitochondria, ribosomes, golgi apparatus and other organelles. But this is only the basic scheme. Not all cells are made equal. And when it comes to neurons, that sketch doesn't quite cover it.

The name 'cell' derives from *cellula* – the Latin word for a room. But the whole thing is built much more like a walled city. The important stuff – the executive decision-making – is done in the nucleus: the town hall. This is where densely packed double-stranded DNA is woven and stored – the blueprints from which your body, and indeed all life, is built. The surrounding clear cytoplasm is dotted with tiny organelles, much smaller than the nucleus, which function like a city's utilities and amenities. Here the mitochondria serve as power stations while ribosomes are industrial estates, given over to the execution of manufacturing orders handed down from the nucleus. Elsewhere in the cytoplasm there are other microscopic structures which play structural roles or take part in waste disposal or defence.

The pyramidal nerve cells of the brain's motor cortex stretch out over vast distances within the body. The extensions of the cell are called axons. For the longest neurons in the body, those axons can grow to be over a metre in length – an enormous distance given the minute scale of the cell itself. To put that into context, consider this: if the cell body of the lower motor neutron were indeed a city, say about the size of London, its axon would be represented by a road that ran out into space for about twenty million miles (which could get you about halfway to Mars!).

The neuron sends its axon down through the brain, on into the brain stem and through the spinal cord, running and converging with others, like individual telephone wires combining to form the main trunk. Most of them eventually cross over to the other side of the body (which is why a stroke on the left side of the brain can lead to paralysis on the right side of the body). And here, in the front of the spinal cord, they end. This nerve cell, the first link in the path from brain to muscle, is called the upper motor neuron. It has so far carried a nerve impulse from the brain to what is essentially a junction box in the spinal cord.

Here, in a location we call the anterior horn, it will form a synapse, connecting with a final neuron, completing the link between the events in the brain that provide the impulse to move and the physical means by which movement is achieved: the contraction of muscle. This second nerve cell, the lower motor neuron, runs from the spinal cord and its axon finishes embedded in the substance of a skeletal muscle.

It is the junction of these two neurons, in the anterior horn, which is vulnerable to attack by the polio virus. If it invades and destroys the cell body of the neuron, then the entire cellular structure, from spinal cord to muscle, dies back too – for good.

The cells of the nervous system are the oldest in your body. In contrast to almost every other cell type in the human body, they lack the ability to divide and self-replicate. Unlike skin cells, which enjoy a hefty turnover, if neurons become irretrievably damaged or die they are not replaced.

To partially compensate for this lack of ability to regenerate, the central nervous system is buried deep within the core of the body, encased within the column of bone that is your spine and protected in the vault of your skull.

Despite this it remains vulnerable, especially in the face of modern threats like motor transport. And of course the armour of the skeleton is no protection against infection.

During an attack of poliomyelitis, many thousands of these lower motor neurons can be lost. And once deprived of their nerve supply, the muscles supplied begin to waste, giving the characteristic appearance of withered limbs that accompanies paralytic polio.

~

When you talk to virologists about viruses, they have a grudging respect for their foe. Incapable of independent existence, viruses rely upon entering more complex cells and hijacking both their metabolic and reproductive machinery. Their genomes are too restricted in information content to allow them to manufacture the means of their own survival. They have only the simplest instruction set – one that allows them to attach to and enter a cell, and trick it into manufacturing further copies of itself.

But these simple structures have the capacity to destroy the host cells they invade and then spread like wildfire – first from cell to cell and then from person to person. And as a consequence viral pandemics are capable of causing death, disease and personal suffering in many millions.

By the time the epidemic in Mauritius had passed, my father's older sister, Angele, was wheelchair bound. His younger sister, Therese, was less fortunate still. In her the polio virus had weakened the muscles responsible for breathing and swallowing. She would go on to die of pneumonia.

~

In 1952 the polio virus arrived in Northern Europe. But the pattern of attack was very different to that seen in Mauritius. Poliomyelitis, the inflammation and destruction of the motor nerves brought about by the polio virus, is also known as 'infantile paralysis' because in earlier epidemics it was almost invariably young children who were most severely affected.

That pattern of attack and disability continued in developing countries like Mauritius but in Europe polio had for some time been confined to small outbreaks because of improved sanitation, so there was little in the way of natural immunity to the virus among the wider community. When the epidemic arrived in Copenhagen in the summer of 1952, the disease ran rife in adults and children alike. And worse still, the manifestation of the disease in adults was far more severe, with a much higher risk of paralysis of the muscles involved in breathing and swallowing. This form of the disease – hitherto rarely seen in polio epidemics – was commonly fatal.

In 1952 there was no drug or vaccine that physicians could set against polio. When outbreaks hit major cities they created tragedies of the grandest proportion. Thousands were infected, with many hundreds left paralysed or dead. Clinicians in general became nihilistic in their attitudes to the disease. Medicine, it seemed, had little or nothing to offer.

But there was a distant hope – that the respiratory system could be supported artificially with ventilators, as a temporary bridge to survival, while the virus ran its course. And for this the world of medicine would turn to the fledgling speciality of anaesthesia.

~

During an interview for a job with a cardiothoracic unit, an anaesthetist was once asked by a rather pompous surgeon what she thought her role was within the surgical team. 'Oh that's easy,' she replied. 'It's like an aircraft. I fly the plane and you do the in-flight entertainment.' Apparently she still got the job.

There's much more to the art of anaesthesia than injecting a drug and making the patient count backwards from ten. Anaesthetists fly human physiology like pilots fly planes. While you're awake and conscious, your physiology is largely under automatic control, just like a passenger airliner on autopilot. The intricacies of your cardiovascular and respiratory systems are held neatly in balance with your kidneys, gut, liver and the enormous complexity of your brain. And your body's autopilot – its system of autonomic control and feedback loops – is pretty good at the job. In health it keeps things running on an even keel, night and day, beat to beat, even when you're asleep. Evolution has allowed thousands of biological processes to be seamlessly integrated and orchestrated under automatic control, so that you can go about your business and do the stuff of conscious thought without having to be bothered by pesky things like stopping to remember to make yourself breathe, or keeping your heart beating with the right rate and force.

But the unconsciousness of anaesthesia is something other than sleep. It's a little bit like rebooting that autopilot mid-flight and handing the aircraft over to someone else for manual control. And in the same way that the captain of the plane takes over control to gently navigate around bad weather, so the anaesthetist must wrest control of physiology from the patient in order to navigate the hazards presented by surgery, injury and disease.

∿

The scale of the 1952 Danish polio epidemic was unprecedented. In Copenhagen over three thousand people were infected, among whom more than a third showed signs of paralysis. And the number of these patients suffering with respiratory failure was higher than in any other European outbreak. Copenhagen boasted several large municipal hospitals but there was only one, the five-hundred-bed Blegdam Hospital, that was equipped to deal with infectious diseases, and it was here that cases began to arrive.

Towards the end of the summer, the polio epidemic was in its fullest throes. Henri Cai Alexander Lassen, Professor of Epidemiology at Blegdam, charted the progress of the outbreak and was shocked by the tidal wave of disease and death that flowed through the hospital's doors. Among the facility's staff there was frank desperation; the disease appeared to defy any conventional treatment. In the first three weeks of August, thirty-one patients suffering with paralysis of the muscles of breathing and swallowing were treated at Blegdam. Despite the hospital's best efforts, all but four died. Desperate for a measure that might turn the tide against the virus, one of Blegdam's physicians, Mogens Bjorneboe, recalled the work of an innovative young doctor called Bjorn Ibsen who was interested in anaesthesia and artificial ventilation. Ibsen was a freelancer among the hospitals in Denmark and Bjorneboe had worked briefly with him earlier that year in treating and ventilating a newborn suffering with tetanus. The child did not survive but the intervention itself appeared, to Bjorneboe, to have worked at least briefly. Ibsen was promptly summoned.

~

Three years earlier, in 1949, Bjorn Ibsen had travelled to Boston to train as an anaesthetist at Massachusetts General Hospital. He spent a year there and returned to Denmark with new skills and insight. He was nothing if not unconventional. He chose anaesthesia over more traditional careers – a brave decision in a world that wasn't yet ready to acknowledge that this was a speciality worthy of the attention of qualified doctors.

He returned to Copenhagen in 1950 to find his former tutors scornful of his experience. The University Hospital of Copenhagen regarded Ibsen's sojourn abroad as though it were time spent in the wilderness. 'You have been away from the fountain of life for one year,' remarked a professor of surgery. 'Let us hope you can catch up with what you have missed.' Despite these verbal assaults Ibsen thought that the anaesthetist might find a role well beyond the walls of the operating theatre. After all, the experience of resuscitating a patient bleeding to death from a brisk haemorrhage or managing the life-threatening side effects of primitive anaesthetic agents gave the anaesthetic fraternity an appreciation of real-time applied physiology that was otherwise lacking in medical practice.

But Ibsen – having witnessed isolated cases of polio and with first-hand experience of the slow suffocating death that it brought – was most interested in the anaesthetist's ability to take over a temporarily compromised organ system.

During the polio epidemic in Copenhagen the most fortunate among the patients were treated with artificial ventilators called iron lungs, which assisted breathing by helping the patient expand their chest. These devices were half-cylindrical vacuum chambers, large enough to accommodate an adult. They were constructed so that a patient could lie with their body sealed inside and only their head protruding through a hole in the top, sealed around the neck with rubber. The pressure inside the cylinder, and therefore inside the patient's lungs, could be reduced to below that of the outside air, creating a partial vacuum in the patient's chest and sucking air into his or her lungs through the mouth and nose. In this way the iron lung devices mimicked the normal mechanism of the lungs, using reduced pressure inside the chest cavity to suck air in from outside. This became known as negative pressure ventilation.

Ibsen realised that iron lungs were effective but cumbersome, expensive and, when it came to the hospitals of Copenhagen, in desperately short supply. Because of this their use was severely rationed, and during the polio outbreaks doctors had the unenviable task of deciding who, among the dozens of victims, should be given this chance of life and who should be left to die. So scarce was the resource that even when the iron lung ventilators were employed, they were often used too late to make a difference.

And while the worst cases of polio in the Copenhagen epidemic were proving almost invariably fatal, Ibsen was nevertheless confident that the skills and knowledge he had acquired while in the United States could save lives. Ibsen believed that the early failures seen at Blegdam were partly attributable to clinicians' poor understanding of both the disease and its effect on human physiology.

No one seemed sure why these patients were dying. The sickest patients were drowsy and febrile, to the point where some of the doctors assumed that polio was causing infection and inflammation of the brain.

But Ibsen disagreed. The drowsiness and rapid heart rate, he believed, were not the result of encephalitis caused by polio but the consequence of high levels of carbon dioxide accumulating in the bloodstream.

~

In addition to bringing fresh oxygen into the body, the lungs are also responsible for expelling carbon dioxide. And while deficient levels of oxygen in the bloodstream can, in part, be treated by increasing the amount of oxygen inhaled, the expulsion of carbon dioxide from the lungs depends much more heavily upon the rate and depth of breathing. Ibsen measured the levels of carbon dioxide in the bloodstreams of the sickest polio patients. Levels of oxygen appeared to be normal in these patients but carbon dioxide had, in contrast, accumulated to many times its normal level.

Artificial ventilation was the answer, Ibsen was sure of it. He had taken great interest in the work of Dr Albert Bower and his colleagues in Los Angeles, who had described their ventilation of polio sufferers with iron lungs and how this had reversed their prognosis from 90% mortality to nearly 80% survival in less than four years. If the Danish polio patients suffering with paralysis of the muscles responsible for breathing and swallowing could be similarly ventilated then perhaps they could hope for the same success rates.

But iron lung machines were bulky and hugely expensive – about the same as the average 1950s family home. And Blegdam Hospital possessed only three.

A cheaper, more widely available alternative would have to be sought. Here Ibsen fell back upon his experience in the operating theatre. He knew that patients could be ventilated by passing a tube into the trachea, connecting a rubber bag to the end of the tube and then allowing oxygen to run into the assembly. When squeezed the bag would push fresh oxygen into the lungs, thereby inflating them. When released the elastic recoil of the lungs expelled air laden with carbon dioxide through a valve. This method of ventilation moved air into the lungs by applying positive pressure from the outside rather than trying to replicate the work performed by the respiratory system in generating negative pressure within the chest. Ibsen was sure that this would work outside of the theatre too. The scheme required little equipment and so could offer a lifeline to dozens of patients, rather than the few who could be serviced by the handful of iron lungs that the hospital possessed. But Ibsen's method would have to be demonstrated and proved before his physician colleagues would accept it. He would not have to wait long for the opportunity.

∾

Just a few days after Ibsen first arrived at Blegdam Hospital, he was referred the case of a twelve-year-old girl whose limbs and chest were paralysed and who could not swallow. Breathless and unable to deal with the saliva in her mouth, she was choking on her own secretions. Her case was near identical to that of the twenty-seven patients who had died in the previous month. Without intervention it seemed certain that she too would die.

Ibsen took her to the operating theatre and persuaded a surgeon to perform a tracheotomy: making a hole in the neck, around an inch below the Adam's apple, which could admit a breathing tube.

The surgery proved difficult. They had injected a local anaesthetic agent into the skin where the incision had been made but the girl was agitated and fought against the medical team. The surgical wound bled back into her airway, soiling her lungs and adding to her distress. By the time tracheotomy was complete, and Ibsen's rubber breathing tube had been inserted through the new opening, she was in extremis, with Ibsen wrestling to retrieve the situation. His colleagues, who had gathered to observe his efforts, assumed that they were merely witnessing the futile efforts of a physician to revive yet another patient dying of polio-myelitis. One by one they turned their backs and left the room.

Ibsen had to think quickly. The girl on the operating table before him was suffocating. The tube connecting her lungs to Ibsen's rubber bag was in place and free of obstruction. But she was now distressed and fighting against Ibsen's efforts to squeeze air into her lungs. With no air entering or leaving her chest the oxygen in her bloodstream was dwindling while carbon dioxide was on the rise. If she was to survive he would have to stop her from fighting against him and take over her breathing completely. Ibsen injected her with sodium thiopental, an anaesthetic agent, and within seconds her body had gone limp. Now, for the first time able to squeeze air into her lungs, Ibsen could make headway. Asleep and unable to resist Ibsen's efforts, she was finally breathing – albeit artificially and with his assistance. The colour returned to her face and, as the carbon dioxide fell, her heart rate stabilised.

Ibsen's physician colleagues returned to the room, incredulous that he had rescued a child who, but a few minutes earlier, had been so clearly at the point of death.

The hospital wasted no time. Ibsen's technique was adopted and within eight days the wards were filled with patients being ventilated using this technique. Armies of medical students and nurses were recruited to assist in the task; standing by bedsides, squeezing bags in shifts, day and night, they provided artificial ventilation to dozens of patients at a time.

∿

Up until the middle of the twentieth century, medicine was mostly about the treatment of chronic illness: consumption, cancer, syphilis, arthritis and the like. Short, severe illness was generally fatal. Survival was rarely attributable to heroic medical intervention. With the exception of a few genuine medical emergencies that could be solved with a knife, there was little that the art of medicine could put in the way of critical illness. The idea that medicine might be in the business of buying the patient time by supporting their vital organs against the onslaught of overwhelming disease, was almost entirely alien. But Ibsen's pioneering work in the field was to have far-reaching consequences. And poliomyelitis was by no means the last viral epidemic to threaten the world.

~

On 11 March 2003, Carlo Urbani was on his way from Hanoi to Bangkok, attempting to relax after what had been a frenetic and exhausting fortnight. Based in Vietnam with the World Health Organisation, Urbani had been called in to advise physicians at the French Hospital in Hanoi on 28 February. There, a Chinese-American businessman named Johnny Chen had been admitted, suffering with an unusual and serious flu-like illness. Urbani was unsure of the identity or nature of the disease; it behaved unlike anything he had seen before. And as Chen's condition deteriorated, Urbani's concern at the strangeness of his illness grew. Within days, members of the medical team who had been in contact with Chen were also falling ill and exhibiting the same constellation of symptoms. It was clear to Urbani that they were dealing with a new and potentially dangerous infectious disease.

Chen, a man in his mid-forties, had a high fever and what looked like severe pneumonia. But other organ systems were also involved: his blood pressure was dropping and his kidneys were showing signs of compromise. The medical team investigated further but none of the usual suspects were present; bacteria were absent from his bloodstream and the course was too aggressive to be ordinary viral influenza in a reasonably young, previously healthy man. The disease was a mystery. It was without a name, a known cause or a point of origin. Without these it would remain without a treatment, a vaccine or a means of containment. And if, as seemed likely, it proved lethal to Chen, then Urbani would be looking at an unknown, fatal and highly infectious disease in a man who had travelled halfway across the world aboard a sealed jet aircraft, making countless contacts on the way.

Reports of a severe and atypical pneumonia sweeping across the Southern Provinces of China had been circulating for some months. But details and reliable data had been frustratingly hard to come by. Chinese officials had initially played down the scale of the outbreak, stating that the number of cases ran to little over three hundred with only five deaths among these. This implied that the mystery illness was of little concern and would most likely burn itself out. But the true extent of the outbreak had been disguised. Later the world would learn that over eight hundred people had become infected in China in those early months and more than thirty had gone on to die. But in February 2003, Urbani and the medical team at the French Hospital in Hanoi knew nothing of this.

Urbani spent the next eleven days working closely with the French Hospital in Hanoi. He first told the staff how best to protect themselves with the equipment they had available. At this time they had little more than gloves, hand basins and medical masks, but Urbani impressed upon them the vital importance of these basic measures. As concern grew among the hospital staff, Urbani provided reassurance through his continued presence. He returned every day and worked late into the night. Through these efforts he built trust and later helped the hospital to take the difficult step of quarantining those members of staff with symptoms away from the wider Hanoi public. Shortly afterwards, the French Hospital closed itself to the public and posted armed guards outside its front doors.

Urbani's instincts told him that this was something very strange and very dangerous – something other than flu. He pursued lines of enquiry relentlessly, working long days at the French Hospital, taking samples, running tests and making sure that infection control protocols were properly enforced. Containment and proper identification of the causative organism were his priorities. The war against this infectious disease, whatever it was, would turn on these simple measures.

Pascale Brudon, the head of the World Health Organisation's regional office in Hanoi, witnessed Urbani's efforts and was in touch with him throughout. Together they saw to it that the WHO's headquarters in Geneva were alerted. If their instincts were correct, then the fallout from this disease would be felt all over the world.

Over the next few days international experts, summoned by the Vietnamese government on Urbani's recommendation, arrived in droves. By this point, Brudon could see that Urbani was exhausted. He had for the past fortnight been alone in the fight to identify and contain this disease and now clearly needed to rest. Brudon suggested that Urbani could now afford to take a break and attend a conference in Bangkok, where he was due to give a lecture. Fatigued, Urbani accepted and on 11 March 2003, after handing over to the incoming teams from the WHO and the United States' Centre for Disease Control (CDC), he boarded a plane at Hanoi airport.

Aboard the flight, Urbani developed a fever, a dry cough and a headache. In those hours, confined aboard that aircraft, he could have been under no illusions about his ailment's likely cause. After the plane touched down, Urbani found his way through to meet a colleague from the CDC who was waiting to greet him. Fearing the worst, Urbani urged him not to approach. While they waited for an ambulance to arrive, the two men sat in silence, separated by a distance of more than eight feet. The paramedic team arrived wearing masks and protective clothing and took Carlo to hospital. He died eighteen days later.

$\backsim$

In the same week that Carlo Urbani left Hanoi, Johnny Chen, the forty-eight-year-old businessman whom Urbani had first been called to see, died in intensive care after having been transferred to Hong Kong.

Days later, Paul Derosier, a sixty-five-year-old French anaesthetist who had treated Chen, along with a nurse who had been involved in his care, also died of the same disease. By 15 March, authorities were aware of forty-three cases in Hanoi. Of these, forty-two were healthcare workers who had looked after Johnny Chen. The exception was the son of one of the infected hospital staff. Among these, five had deteriorated rapidly and been taken to the hospital's intensive-care unit for artificial ventilation.

The WHO had also become aware of new cases worldwide in Singapore, Taiwan, Canada and Hong Kong. In the week that Carlo Urbani was himself admitted to an intensive-care unit in Bangkok, the WHO issued a global health warning for the first time in its fifty-year history. And the disease, whose precise nature was still a mystery, would finally get a name: Severe Acute Respiratory Syndrome or SARS.

By the time of the WHO's health warning this much was known: the disease was infectious, highly transmissible and deadly. Health workers on the frontline and their families were most at risk.

Due in large part to the efforts of Urbani in the early days of the outbreak, the origins of SARS were rapidly established. It emerged that Johnny Chen had travelled from Hong Kong; there he had stayed on the ninth floor of the Metropole Hotel. Here he and seventeen other guests appear to have acquired SARS from a single individual. Dr Liu Jian-Lun, a sixty-four-year-old Chinese medical professor, had unknowingly contracted SARS in Guangdong while treating patients. He had travelled to Hong Kong to attend his niece's wedding. This journey from the South-Eastern Provinces of China to Hong Kong was the triggering event in the global outbreak that followed. Room 911, the room occupied by Dr Jian-Lun, became the centrepiece of the investigation. And the ninth floor of the Metropole became ground zero for SARS.

The virus itself had circulated in animals for many months. Virologists chased its origins back to civet cats. In the food markets of Guangdong, with their exotic animal husbandry, it had moved from animal species to animal species before finally making the jump into humans.

Precisely how it did this remains a fundamental question for the science community. The limited repertoire of genes that the virus possesses is able to mutate and re-assort. It is like the badly copied blueprint for a curious device, handed down from one generation to the next. Offspring are able to share new innovations or spontaneously improvise, until finally enough of those alterations align and sum to produce a terrible weapon. Nature, as our virologists are fond of reminding us, is the best and most efficient bioterrorist.

But SARS would have likely remained endemic within the Southern Provinces of China had it not been for the fateful journey of Dr Liu Jian-Lun. Taking it to Hong Kong, to an international business hotel, provided a most efficient vehicle for the spread of disease. At that nodal point, Jian-Lun was confined and in contact with dozens of travellers, all of them passing through, many on their way to other international destinations. From the moment Jian-Lun checked in to the Metropole Hotel, SARS was set to go global.

∾

SARS, as its name suggests, first affects the respiratory system. But unlike polio it does not target the mechanics of breathing but rather the substance of the lung itself. Cells in the tissues of its fragile air sacs and the branching network of airways become bound by virus. The virus enters and forces these cells to start churning out millions of new copies, like a printing press turned over to the production of quick and dirty war propaganda. The cells are not entirely without response. They are able to signal that they are compromised and summon the immune system to attack. But the virus is buried deep within the structure of the cell, and so destroying it means destroying the cell in its entirety: collateral damage in the wider fight against disease.

A combination of the death of these infected cells and the scarring and inflammation that accompanies the immune system's attack leaves the substance of the lungs compromised. The once tissue-thin membranes, capable of expanding and collapsing likc a supple balloon, become more rigid and less compliant. The exchange of oxygen and carbon dioxide across their surfaces is obstructed and the force needed to expand the chest and perform the work of breathing is massively increased.

To the physician called to see a patient deteriorating in the face of SARS, the signs are all too clear. Effortless healthy breathing is replaced by a rapid, shallow pattern. Other muscles not usually involved in expanding the chest are recruited to overcome the stiffness brought by the viral infection. All of this additional mechanical effort needs to be paid for. And so the body's demand for oxygen increases at the same time as its ability to grab those molecules of oxygen from the outside air, and exchange them over the thickened, diseased membranes of the lung, worsens.

Haemoglobin, the molecule in the blood cells that carries oxygen, is bright red in appearance when fully laden. Once stripped of this oxygen load it becomes duller and bluer – accounting for the difference in appearance between arterial and venous blood. But if arterial blood cannot acquire a new, full load of oxygen in the lungs it loses its rosy hue. The skin through whose capillaries these blood cells course acquires a shade more akin to thundercloud grey.

It is that vision – of the grey, breathless patient with the thousand-yard stare – whose first glimpse, even in the half light of a hospital ward at night, signals real trouble and the need for interventions that can only be provided by intensive care. When the supply of oxygen is outstripped by demand, critical illness and death will inexorably follow. In these circumstances, the bridge to survival is provided by modern intensive care.

$\sim$

Charles Gomersall was at the end of another shift as consultant in charge of the intensive-care unit at the Prince of Wales

Hospital in Hong Kong. He had worn his hard shell mask all day and its metal pinch clip had dug itself into his face, leaving a reddened dent in the bridge of his nose. But even now, striding across the car park, away from the ward and main hospital building, he kept it in place. The past fortnight had been punishing. The SARS outbreak was now at its height and the unit was under strain from the constant flow of cases in need of critical care.

His first week on duty during the epidemic had been sobering. As an experienced intensivist he was familiar with destructive pneumonias and deranged physiologies, and used to holding the line in the face of adversity, but SARS had a different character. The clinical course was so fierce that at first Gomersall wondered if any of his infected patients would manage to survive.

The damage to the respiratory system wrought by the virus was severe. Artificial ventilation had to be applied with care. Forcing stiffened lungs open with external pressure from a ventilator was not without its hazards. Titrating the volumes and pressures applied by the mechanical ventilators precisely against the needs of each individual patient was an art. Getting it wrong could rupture delicate membranes, causing pneumothorax and a life-threatening collapse of the lungs. And ventilating too hard, with overzealous volumes, could further inflame the lungs and the situation. But it was the impact of SARS upon the rest of the body that presented the biggest challenge.

~

The cells of the immune system roam the bloodstream and tissues, like policemen pacing the beat. They detect potentially harmful microbes, attack them and then beckon other immune cells to enter the fray. This system of self-amplification is like calling for backup at the scene of a jewellery heist. When activated appropriately, it puts a stop to trouble before it has a chance to get out of hand. But this system can be all too responsive. Some infections – SARS among them – over-recruit the immune response, giving rise to widespread inflammation which in turn can harm the body. It's like trying to stop a carload of jewel thieves with the unrestrained might of the nation's combined armed forces: an inappropriate response that causes more damage than the crime itself ever could have.

Because of this, Gomersall's SARS patients endured more than simple respiratory failure. The storm of immune response damaged kidneys and livers and caused hearts to fail. And this multi-organ failure also had to be supported.

In the years since Ibsen's first intensive-care unit was established, medical technology has moved on to allow the carefully nuanced support of many organs beyond the lungs.

Now the failing circulation can be supported with noradrenaline, which raises sagging blood pressure. The heart can be driven with infusions of adrenaline, boosting its contractile force and ejecting greater volumes of blood needed to perfuse the rest of the body. And medicine has learnt how to replace the work of the kidneys, using dialysis machines and blood filters. Even a malfunctioning gut can be augmented with a feeding tube or replaced by running calories and nutrients directly into veins. Today all of this can be achieved artificially and, in the most dangerous days of the disease, with patients in a state of anaesthesia and unaware of their plight.

But Gomersall, along with other doctors and nurses on the intensive-care unit, was fast becoming fatigued. It was unheard of to have so many patients dependent on such high levels of artificial support for such a prolonged period of time. At this point the outbreak had been raging for weeks and there was no end in sight. What's more, SARS was threatening the lives of the very front-line medical professionals who were struggling to keep its victims alive.

Protecting the clinical team had become a priority, and one that Gomersall's intensive-care unit had found itself initially ill prepared for. The high filtration masks, so essential to prevent droplets laden with virus from penetrating into the healthcare workers' respiratory tracts, were in short supply. And even once these arrived, it turned out that they needed to be tested for a precision fit: a poorly fitting mask was worse than no mask at all. This procedure could take up to twenty minutes for each person – a frustrating delay in the middle of the frantic battle against death and disease.

There were other, unanticipated problems. It was the beginning of the Hong Kong summer. Ambient temperatures ran at close to 30°C with humidity at nearly 80%. The personal protection equipment covered the ICU team members from head to toe, leaving only a few square inches of skin exposed. The heat stress was stifling, even with the unit air-conditioning set to a usually bone-chilling 17°C. But despite fastidious attempts to avoid infection, the intensive-care staff found that even their cumbersome masks, gloves and protective clothing couldn't keep them safe from SARS. In all, five of their team contracted the disease and one was later admitted to the intensive-care unit. But despite the dangers to themselves and their families, the doctors and nurses of the Prince of Wales Intensive Care Unit continued to show up for work, week in, week out.

Gomersall got into a routine. As soon as the severity of the situation became clear he moved out of the family home, away from his wife Carolyn and two young daughters. He rented a flat nearer the hospital and travelled to work by car. The act of getting in and out of the protective garb, to eat, drink or go to the toilet, was time-consuming and left him vulnerable to infection. And so Gomersall took to waking early in the morning to breakfast and take on a decent load of water to hydrate himself. He then worked through the day without having to get undressed or remove his mask. Only when safely back inside his own car did he finally take the mask from his face. Each day, when he got back to his flat and closed the door, he felt a sense of overwhelming relief to be away from the ward and in his own space again. There, alone, he was in no danger of infection. More importantly still, he was at no risk of passing the virus on to anyone else.

Gomersall would work for five days in a row on the unit. Before he could go home he had to make sure that he wasn't incubating SARS. To do this he would spend ten days away from the ward, teaching and doing administrative tasks in his office – still staying at the flat. At the end of that time, if he wasn't sick and hadn't developed a fever, it would be safe to go back to his family. Gomersall went through this cycle of work, self-imposed quarantine and brief family reunion three times.

He got only four days at home between each shift. His family would studiously avoid talking about the elephant in the room. SARS dominated the news. Hong Kong had been paralysed by it. But Charles didn't much want to talk about what he'd seen, and his wife Carolyn didn't want to hear about it. Should he fall ill at work, Charles had told Carolyn that she should not come and visit. To lose one parent to SARS would be tragic; to lose two – as some families in China already had – would be insupportable.

Every day the teams faced the same set of problems: an intensive-care unit full of people ravaged by SARS, hopelessly unwell, propped up by a constellation of machines and drugs. And in truth these weren't much more than a way of buying time in the hope that the disease would abate. That is all intensive care ever is: an extraordinary effort on the part of medicine to stretch human physiology well beyond its survivable limits in the hope that the patient can stay alive until something changes for the better.

In mid June 2003 something did change. For the first time since the SARS epidemic began, no new cases were being admitted from outside the hospital. The only infections now were happening on the wards, between patients and healthcare staff. SARS, for all its ferocity, had a peculiar pattern of behaviour that had limited its spread. Some viruses, influenza for example, are highly transmissible from very early in the infection, long before the patient becomes incapacitated and unwell. This is why flu spreads so quickly and so widely. Many people infected with flu remain well enough to go about their business, shedding the virus to the outside world all the while.

But in most cases of SARS, the peak of contagiousness occurs only once the victim has become critically unwell, usually in the second week after infection. By this time most of the patients had already been admitted to hospital. This was why healthcare staff had been so badly affected. And though the virus was both highly transmissible and deadly at this point, this limited the spread of SARS in the world outside the hospitals. By mid July 2003, a little over four months after Carlo Urbani had first been called to the French Hospital in Hanoi, the SARS outbreak was firmly in decline and the last of the travel restrictions to affected areas, recommended by the WHO, were lifted. Worldwide there had been more than eight thousand cases with 916 deaths among these. By the following May, no new cases were being reported to the World Health Organisation. The chain of spread, from human to human, had finally been broken.

It could have been far worse. Carlo Urbani's heroic efforts in the early identification of the disease, and his swift actions in notifying the World Health Organisation's Headquarters in Geneva, led to a series of events that contained outbreaks and limited the overall spread of the disease.

Urbani first reported his concerns in early March 2003. After tracking its rapid dissemination to three other countries, the World Health Organisation issued its global warnings a fortnight later. Before the month was out, Malik Peiris' laboratory at the University of Hong Kong had identified a new coronavirus, SARS-CoV, as the probable causative agent, and within a month of that a Canadian laboratory succeeded in sequencing its genome. This provided information vital to the development of diagnostic tests and vaccines. But with travel to affected areas restricted and quarantine measures in place, the virus burnt itself out.

The fight against epidemics and global pandemics is won not by high-tech interventions but by public health measures. In this context, the work of intensive-care units may appear as little more than a gesture: the symbolic fighting of brush fires in a world under threat of being engulfed by a massive conflagration.

Indeed, the polio epidemic, which gave birth to the speciality of intensive care, was defeated not by ventilators, adrenaline pumps or dialysis machines but by a programme of vaccination – a campaign so effective that today the polio virus stands on the brink of eradication from the world. Since then, intensive care has re-tooled and re-purposed itself. But the question remains: what is the value of intensive-care medicine – a speciality that invests so many resources for such marginal gains in the face of critical disease?

We can reassure ourselves that it is far more than just a futile gesture. Of the sickest patients admitted to intensive-care units during the SARS epidemic, three out of four survived. Without the battery of artificial support, none would have lived. And mortalities in the worst-afflicted patients of Copenhagen's polio epidemic of 1952 fell from 90% to less than 20% as soon as Ibsen's innovations were implemented.

Today intensive care is a branch of medicine that allows other specialities to undertake more ambitious surgeries and interventions than ever before, safe in the knowledge that intensivists have successfully redefined the limits of human life when challenged by disease and injury.

At times of great crisis, the polio and SARS outbreaks included, intensive care has been to medicine what it is to the individual today: a temporary but much-needed bulwark against illness, a means of buying precious time.

~

We arrive in theatre and shock again before the surgeons begin. The ventilator is running. The patient's lungs too are now beginning to fail – becoming stiffer and demanding more oxygen. The acidosis in his bloodstream is worsening and his kidneys are deteriorating. We increase the adrenaline and the noradrenaline. The doses are now so high that their side effects are becoming a real problem. The drugs make his heart more irritable, more prone to fatal arrhythmias. We can hold his blood pressure up but we're having to defibrillate more often now. Each time the ECG flips into a shockable rhythm, the defibrillator spits out an inch-wide strip of paper on which the jagged trace is printed in hard copy, like a seismograph beating out the lines of an earthquake. Several feet of this strip have now collected on the floor. An alarm goes off. I nod at the surgeons. They step back from the table. We fire the defibrillator again.

This is absurd. Sooner or later the rhythm of his heart will degenerate into something we can't treat, something that electricity can't reset. Perhaps, realistically, that is all we're waiting for.

But then the surgeons call out. They've found a section of dead bowel, its arcade of vessels blocked by something – a blood clot perhaps. Deftly, the surgeons snip out the gangrenous tissue and join healthy ends of bowel together. Things do not change immediately but with the diseased bowel gone and no longer leaking toxins into the circulation, my patient's physiology will get better rather than worse. Surviving the next few days will be no mean feat but the surgeons have given us the means to turn the corner. They are the change that we have been hoping for. We are far from out of the woods, but at least the woods are no longer on fire.

Back on the intensive-care unit, in the hours after the operation, the support we need to provide steadily decreases. We still deliver shocks, but they are fewer in number and less frequent. And slowly the patient is weaned off the drugs and the artificial ventilator. Over the next few days we gradually hand control back to the patient, shutting off our machines as his normal physiology reasserts itself. Precisely how his body is able to recover and knit itself back together after such an insult is unclear. But he is young, and the young are remarkably resilient. And less than four weeks later, that eighteen-year-old walks out of the hospital.

CHAPTER 6
WATER

November 2011: the author diving in the Red Sea using a vintage scuba rig of the type first seen in the 1950s

QALITO WAS THE name of the island. It was part of a Pacific Island chain, fringed with the sort of beaches they use to advertise credit cards, small enough that you could walk around it in just over an hour. The surrounding waters were crystal clear and warm enough to dive without wetsuits. We were part of a coral conservation expedition, based in the Mamanuca island group just a few miles north of Fiji. I had arrived a month or so before Christmas 2003, taking up a less-than-onerous post as dive medical officer. We were cut off from the rest of the island, living out of an abandoned house set back from the beach. For electricity we depended on a diesel generator. Water was delivered by ship and stored in a huge concrete reservoir to the rear.

The house was large but crowded. Bunks filled with volunteer conservationists, scientists and dive instructors lined all of the available space. The volunteers – mostly gap year and university students with a light sprinkling of more seasoned adventurers – rotated in and out for weeks or months at a time. I slept on a mattress on the beach beneath a palm tree with a mosquito net draped over one of its branches. I'd turn in each night shortly after they flicked the generator off and wake when the Sun rose above the line of hills on the island opposite ours. As far as expeditions went, you'd have trouble finding a location less extreme.

177

We handed out advice about proper hydration and sun cream to the newcomers. In every rotation there was always a small cadre of Brits who didn't heed the advice, choosing to loll around on the first day wearing neither sunblock nor T-shirt – and then the next week wincing whenever anything touched their skin or they caught sight of themselves in the mirror. Sunburn, cuts, grazes and hangovers were our staple caseload. The biggest challenge was stopping the gap year guys killing themselves with duty-free alcohol on Saturday nights.

I swotted up on the local marine hazards. I learnt the following: moray eels have bad eyesight but powerful jaws. The venom of the banded coral snake is among the most potent of any snake in the world. The four most dangerous sharks are the great white, the oceanic white tip, the bull and the tiger. But mainly I learnt that in the water you only really need to worry about the water itself.

~

'Brace! Brace! Brace!' he shouts, running the words together as though they were one. I shove my head up against the wall of the helicopter, my crash helmet clunking against the bulkhead, and fold my arms over my chest, hooking my thumbs under the shoulder straps of the four-point harness. We hit the water in darkness and immediately begin to sink. The water is already at my ankles. I rest my right hand on my harness's quick release and with my left I find the handle that will jettison the window. Once under water the helicopter will sink a metre every second.

I never used to understand how it could be difficult to escape from a sinking vehicle. Open the door, swim out and up to the surface. How much of a challenge could that be? On dry land I can hold my breath for the best part of three minutes and I'm an OK swimmer. How long could it conceivably take for me to get to the surface from, say, twenty metres down? But of course you have to factor in the harsh realities of the physics and physiology of your predicament. How long could it take? Very probably forever.

This is HUET (pronounced *hew-it*), the Royal Navy's Helicopter Underwater Escape Training facility in Yeovilton. It exists to provide helicopter crew with the training they need to escape a vehicle that has ditched in open water. The work they do is vital. For more than 80% of helicopter crashes over water, the time between warning and impact is less than fifteen seconds. Of these more than 70% sink immediately with over half of them inverting. The military's experience of helicopter accidents into water is also pretty sobering. Of those occurring in daylight, the survival rate is 88%. But for survivable helicopter crashes into water occurring at night that number is as low as 53%.

But why is this happening? These are healthy people, trained military personnel, and in most cases strong swimmers. The answer lies in the very structure of our bodies.

~

We take our natural buoyancy for granted – mainly because the vast majority of us never dive beyond the point at which we are more likely to sink than float. From the surface, for the first seven metres or so, it takes a bit of effort to dive below the waves. The air in your body, principally that in your lungs, serves as a kind of float and acts to keep you buoyant. Here the upthrust you experience by virtue of good old Archimedes' principle is more than enough to return you from whence you came.

But below those few metres the relationship is reversed. Your tissues become compressed, the volume of air in your lungs decreases as the pressure mounts and you eventually become denser than the water around you: an object that would rather descend into the depths than float towards the safety of the surface.

This state is described as 'negative buoyancy'. It's a strange term when you think about it – like referring to the state of being poor as being 'negatively rich'. What we're really talking about is sinking, and sinking as a probable prelude to drowning.

∿

The water is rising fast now, already up to my waist, and every fibre of my body is telling me that I should unclip that harness and punch through that window. But to do that would be fatal. Free of the seat I'd be swilled around the cabin by the inrush of water; finding my way to the exit, and then locating the metal bar that jettisons the window, would be impossible. If I'm to survive this, I have to wait. The water continues to bubble into the cabin. It's at my chest now and the whole vehicle is overbalancing, biased by the weight of the engines and rotors above, turning upside down in the darkness. The water is up to my chin as the cabin starts to rotate. These are my last few breaths and still I'm strapped into my seat, resisting the urge to get the hell out of there.

~

Holding your breath: that's what your survival boils down to here. It is, on the face of it, a simple act of mind over matter, a discipline you should be able to find within yourself – especially if your life depends upon it.

And yet the desire to breathe is among our most primitive urges. We're designed to draw air into our lungs, to exchange fresh oxygen for the waste gas of carbon dioxide. Our lives depend upon this perpetual toing and froing of gases and it is worth taking a moment here to consider your respiratory system in all its glory.

When we describe the path that oxygen takes from the outside world to its final destination in our mitochondria, we do so as though it has agency of its own. We talk of molecules of oxygen moving into our bodies, diffusing across membranes, arriving at mitochondria, almost as though they know where they want to go. But of course oxygen has no free will of its own. In the act of living your body must solve the problem of how to grab molecules of this gas from the atmosphere and bundle them into cells in sufficient concentration that they can do the stuff of life.

The first part of that performance is the act of breathing. Your ribs are attached to your breastbone at the front and the bony column that is your spine at the rear. At the end of each exhalation they slope steeply downwards towards the ground. Contracting the muscles in the chest wall that do the work of breathing lifts the ribs up, to a nearly horizontal position, increasing the volume of the chest. At the same time your diaphragm, the large dome-shaped muscle that separates the chest from the contents of your abdomen, contracts and drops down, further increasing the volume of the cavity inside your chest.

Your lungs sit inside the cage formed by your ribs, adherent to the chest wall. As the chest moves, your lungs move with them. As the volume in your chest cavity increases, so too does that inside your lungs. The increase in volume leads to a decrease in pressure in your chest. That in turn produces suction, in exactly the same way as separating the handles on a pair of bellows does, and air begins to flow.

That air passes through your upper airways, the larynx and the trachea, and then down into your bronchial tree. I always thought of that branching network of airways as inverted sprigs of broccoli rather than trees. In terms of morphology that's not far off. There's a hollow central trunk which sprouts branches of ever-decreasing calibre, at the very end of which are sac-like structures called alveoli: the buds, if you like, at the end of that sprig of broccoli. The cadaveric lung, formalin-soaked in the medical school's dissecting rooms, is solid and heavy; its airspaces are occupied by pungent preservative fluid. But in life air-filled lungs are lighter than sponge, light enough to float on water.

The anaesthetist's perspective in theatre, watching, as we do, the chest lain open during cardiothoracic surgery, gives a much truer impression of those organs. Watching as they expand and collapse with the rhythmic grind of the ventilator, you are immediately aware of a structure whose volume is principally air, a delicate organ horribly vulnerable to injury.

That fine structure exists to provide a massive surface area over which air can be brought into contact with blood. The alveoli, those tiny air sacs at the end of the bronchial tree, are each no more than a fraction of a millimetre in diameter but each lung holds one and a half million. If you were to unfurl them, and lay them out flat, they would form a mat of tissue half the size of a tennis court at Wimbledon. And that vast area is required to bring enough air into contact with enough blood to keep you alive.

Over the surfaces of those alveoli runs a spider-like network of capillaries, vessels with walls a single cell thick, providing just enough structure to confine the blood cells squeezing through them, while offering the minimum of obstruction to the molecules of oxygen diffusing across their walls.

This is the most delicate interface in your body. Nowhere else is the point of contact between your body and the material from the outside world more insubstantial or delicate. That is why it is buried deep in your chest and protected with a formidable cage of ribs. There is no choice other than to make it that way; it has to be that extensive, that fragile, or else gas would not flow and exchange.

In the act of drowning, that air is replaced by volumes of water, swamping the gossamer thin tissues designed to allow gas to pass from the air into our bloodstream. For the average adult, a total of around a litre and a half of water drawn into the lungs is lethal. A fact all too vivid in my mind, while I sit trapped in the sinking helicopter.

$\sim$

'Wait until all violent motion has ceased.' The words of my training instructor come back to me. What he really means is wait until you're under and upside down; wait until you're really sinking.

I can feel the wetness creeping under my chin and the coldness of the water, and start to take deep gasping breaths. I tell myself it's because I'm trying to drop my carbon dioxide, extending the time before my body senses its levels building in me and thereby lengthening the time for which I can hold my breath. But in truth there are other reasons for my hyperventilation here. The water is at my lips. I tilt my head back and take a last long breath. And then we're under.

It is quieter here, somehow immediately less stressful. When my head was above the surface there was noise and uncertainty.

At least now you know that your race to escape can begin. You can't be sure how long the thing took to invert, how far below the surface you might already be. This is part of the problem. With a sink rate of a metre a second, if it takes longer than seven seconds to get out of the vehicle then you'll be negatively buoyant on your exit. And if it's night and you have no source of light to guide you, the question becomes this: 'Which way do I swim?'

There are stories from those who have escaped from sinking aircraft at night, of people swimming through inky blackness for what seems like an eternity, knowing that if they have it wrong, if they've headed the wrong way, they will swim for the rest of their lives.

Underwater now, the familiar burning desire to breathe is already upon me. But I'm following the instructions. I stretch out my right arm, and the short lever that ejects the window presses into my palm. The black and yellow stripes were the last thing I saw clearly before the wash of water covered my head. Frantically I pull the lever to open the window, then hold the frame to make sure I'm ready to haul my body out from the watery coffin. Only then do I dare undo the seat belt, fiddling with the rotating catch, hoping that it won't jam.

The release comes apart nicely and I pull on the window frame, remembering which way is up. I haul myself out of the helicopter and begin my ascent to the surface. I break the surface with a gasp. It has taken nearly thirty seconds to escape. That's fine in the comfort of this warm swimming pool and simulator. But out there in the North Sea even that's probably going to be too long to save my life. To understand why, we need to think about what makes us breathe.

∾

The act of breathing is one of the few bodily functions whose control is part automatic and part voluntary. You might think that such a vital system would be better off left under the permanent supervision of your autopilot.

The only other comparable rhythm in your life is that of your heartbeat, and that is almost exclusively under automatic control. Yes, you might be proud of your ability to chill yourself out and slow it down a little, but when was the last time you had a game of 'who can stop their heart the longest'?

You can choose the way you breathe; you can choose right now to breathe harder and faster. You can choose to stop breathing altogether. But your body knows what you're like and it doesn't completely trust you. It allows you to take control of your breathing temporarily, but never long enough that you could do yourself any permanent harm. It's not possible to stop breathing long enough to kill yourself; in fact on dry land it's tricky to hold your breath to the point of unconsciousness.

Sooner or later your body and its automatic system of management wrests control from you. And detecting when enough's enough is a bit of an art. Your body is pretty conservative, with a set of early-warning systems that trigger breathing long before your biochemistry gets too upset.

To know that you're not breathing, the body has to detect the way that your physiology changes when gas exchange stops. Broadly speaking there are two things you could detect. Although the falling level of oxygen in your blood would be the obvious thing to use as an indicator of danger, that's not what your body does.

That leaves carbon dioxide. When you stop breathing, the level of carbon dioxide in your blood rises faster than the level of oxygen falls. That means that a high concentration of carbon dioxide becomes a telltale sign that you need to take a breath.

So in part, the early-warning system functions by sensing the effect that rising carbon dioxide levels have in the body. Carbon dioxide molecules dissolve in water and make it more acidic. It's this acidity that indirectly tells the body that the mechanics of breathing have been halted for too long.

But the system is more complicated than that. In fact, nobody is sure precisely what triggers the break point in our drive to breathe: the point at which the urge to take a breath becomes irresistible. We know that there is a constant central rhythm, beaten out by respiratory centres deep in the brain stem – clusters of cells which keep time in a dance that lasts throughout the whole of our lives and one from which we may only very briefly absent ourselves. This rhythm is like a perpetual biological metronome. When the performance of breathing halts, it keeps ticking away, urging you to restart.

And then there are environmental factors. In swimming pool tests – with the water at 25°C – experimental subjects simulating escape from a helicopter could hold their breath on average for no more than thirty-seven seconds. Away from the warmth of an indoor pool, the situation deteriorates further.

In water below 12°C the cold shock response is activated. This, a reflex triggered by the widespread activation of cold receptors in the skin, provokes an involuntary and uncontrollable gasp, forcing an individual to draw huge volumes into their lungs whether immersed or not. The drive to breathe it produces is so profound that the average breath-hold time, in cold water conditions, falls to just six seconds.

We also know that there are receptors that sense mechanical stretch in the lungs and the muscles that do the work of breathing. But what weight each of these indicators carry, how they interplay, and to what extent the mechanisms vary from one individual to another, remain matters of educated guesswork.

It is interesting that in this age we talk about the search for a Grand Unified Theory of physics but have still to achieve a well-unified theory of that most fundamental of physiological functions: what it is that makes us breathe.

~

Giorgos Haggi Statti was, at first sight, an unremarkable Italian fisherman. He was of diminutive height with skin darkened by days spent labouring under Mediterranean skies. His build was slight; his pulse and rate of respiration were regular. His heart sounds too were normal. These details were recorded by an Italian physician, searching for something out of the ordinary that could explain Statti's extraordinary feat. In 1913 Statti became something of a local celebrity, after diving to recover the anchor of a warship lost in the harbour depths. Holding a single breath of air, they say he reached a depth in excess of seventy metres.

Skin diving itself wasn't anything new. The exploits of the Japanese *Ama* had been known for centuries: these remarkable female divers remained submerged in coastal waters for minutes at a time collecting shellfish, sea cucumbers and pearls. However, nothing in the history of the *Ama* suggests that they dived much beyond a depth of twenty metres – certainly nothing of comparable depth to Statti.

Intriguingly, Statti claimed that he was capable of even deeper dives. But nothing about the Italian gave a clue as to how he was able to achieve such feats. When tested on land he could only hold his breath for forty seconds. Only one feature of his physique stood out: a seemingly overinflated, barrel-shaped chest. The only other evidence of his exposure to the great pressure of depth was his impaired hearing. One eardrum was entirely absent and the other damaged and perforated. Statti was bemused by the interest in his salvage dive. To him it was nothing out of the ordinary: he was a fisherman and this was just something he knew that he was capable of. He was baffled by the doctor's questions, and annoyed at being asked to hold his breath on dry land. To Statti it was a meaningless test; for him everything was different once he was in the water.

~

In 1968 a physiologist called Albert B. Craig, working at Rochester Hospital, wrote a paper reviewing the predicted physiological limits for breath-hold diving. With the body of knowledge in hand, it appeared that human beings shouldn't be able to free-dive beyond around thirty-four metres from the surface. This, scientists had estimated, was the depth below which the increased pressure of the surrounding water would crush the lungs and reduce their volume to the point at which blood would pour into their airspaces.

A theoretical limit on the length of time for which a person could hold their breath had also been set at around three minutes. This wasn't just an arbitrary number. Physiologists had calculated it based upon the amount of oxygen left in the lungs of a person of average build after maximal inspiration, and balanced that against the resting rate of oxygen consumption by the body. When the graphs were drawn and extrapolated, the scientists saw that the levels of oxygen in the bloodstream would fall to a point at which unconsciousness was inevitable in less than the time it would take to boil an egg.

And yet Craig was aware of many cases of divers who had dived well beyond those depths and times. And so while science had drawn neat lines in laboratories, delineating the theoretical limits of survival, breath-hold divers had busied themselves swimming right past them. It's not clear who this delighted more: the scientists or the divers.

By the mid 1980s, the record breath-hold dive stood at 100 metres. Neoprene, rubber fins and eye goggles had lent this modern sport of free-diving a new dimension, but among physiologists it remained poorly understood. In a series of international symposia, the dangers of free-diving were hotly debated: the plummeting oxygen reserves and climbing carbon dioxide levels; the theorised volume changes and mounting pressures. But of one thing scientists were sure. There was, somewhere, a depth beyond which breath-hold divers would not survive. As Craig himself said, when queried on the subject, 'I think the limit will be reached when one of these breath-hold divers comes up coughing blood.'

～

The key to these seemingly superhuman performances lies in the mammalian diving reflex. When under the water, nerve endings in the skin of your face sense that you are immersed and kick off a series of adaptive responses which include slowing of the heart and constriction of the blood vessels in your peripheral circulation.

This restricts blood supply to non-essential organs and tissue beds, reducing their oxygen demand and conserving the supply for the heart and brain. This constriction of the blood vessels is in a different league to the gentle blanching of the skin that might, for example, happen when you arrive at the airport check-in desk only to realise that your passport is still at home. In the diving reflex it appears to be a hard clamping down of much larger vessels, among them important arterioles and veins, and this in turn leads to a massive increase in blood pressure.

We measure blood pressure in millimetres of mercury. Assuming that you're fit and well, your peak blood pressure is equivalent to a column of mercury probably no more than 120mm in height. A general practitioner might think about putting you on medication to control your blood pressure if it exceeded 140mmHg. In an accident and emergency department, a reading above 180mmHg might warrant immediate intervention. But for otherwise healthy breath-hold divers, pressures above 230mmHg have regularly been recorded.

These huge pressures reflect the centralisation of blood that happens as the peripheral circulation shuts down. And the rush of blood to the body's central compartment has further benefits. It is thought to engorge the tissues of the chest with blood, allowing airspaces to be compressed beyond the physiologists' theoretically defined limits while at the same time protecting them from damage.

This selective diversion of blood is just one measure that reduces global oxygen demand while increasing oxygen supply to the heart and brain. Another mechanism relies upon the spleen: a fist-sized organ which sits just below the ninth rib on the lower left side of the chest. The spleen can function as a reservoir, sequestering a pool of red blood cells – like the tin of money you might tuck away on a kitchen shelf for a rainy day. The drop in oxygen levels that accompanies breath-hold diving forces the spleen to contract, spilling its supply of 'rainy day' blood cells back into the circulation and, in theory, further increasing the supply of available oxygen.

And having made more oxygen available for vital organs through these mechanisms, the brain sends signals to the heart instructing it to reduce its rate, thereby further reducing demand. The heart is among the most metabolically active tissues in the body. Reducing its rate of work reduces its own oxygen consumption. In free-divers, heart rates as low as twenty beats per minute have been recorded during deep dives.

All of this is an elaborate scheme for managing supply and demand under extraordinarily challenging conditions. These in part explain why the current free-diving depth record stands at 214 metres (held by Austrian, Herbert Nitsch) with the longest breath-hold a staggering 11 minutes and 35 seconds (held by Frenchman, Stephan Misfud).

But these extreme feats do not come without risk or penalty. As the circulation shuts down at the edges and becomes more sluggish, muscles and other less vital organ systems are forced to work largely without oxygen. This anaerobic respiration causes waste products to build up in the bloodstream, steadily acidifying it. This is like the body taking a high-interest, short-term loan from its metabolic bank. In fact it's more like borrowing from the worst kind of loan shark. Except in this instance, failure to promptly repay the debt *always* leads to death.

The oxygen debt is called in at the end of the dive. After surfacing divers take long, gasping breaths which serve to replenish their plundered reserves and rebalance their physiology. For those who misjudge the dive, however, oxygen levels will fall to the point at which consciousness can no longer be maintained. If this occurs, with no one present to immediately assist the diver and return them to the surface, death by drowning is inevitable.

But those who free-dive regularly appear to undergo a series of adaptive changes that seem to reduce the risk of fatal misadventure. The dive reflex becomes more pronounced with training: experienced breath-hold divers exhibit lower heart rates and higher blood pressures on immersion. The tolerance for accumulating carbon dioxide also improves as the bodies of these divers reset their expectations. The cells that detect changes in carbon dioxide levels in the bloodstream respond more laconically, sending signals to the brain less urgently when breathing stops. Lungs become more compliant and their volumes also increase. This, as the doctor examining Giorgios Statti in 1913 noted, leads to a more expanded chest. And most impressive of all, experienced free-divers train themselves to resist the reflexive and primal urge to breathe – a remarkable feat of mind over matter allowing them to push human physiology well beyond its default limits.

$\sim$

Until the middle of the twentieth century, diving was a pursuit that demanded umbilical connection to the surface: a tube through which the atmosphere above could be pumped down to a diver. The Siebe Gorman standard diving dress – a helmet of brass bolted to a waterproofed canvas suit, given ballast by shoes and weights made of lead – became the iconic image of the diving industry. The helmet was supplied with air via a hose from the surface, driven first by hand pumps and later by diesel motors. The link with the surface could not be broken.

Later, self-contained underwater breathing systems were invented, which allowed the diver to swim without being encumbered by an umbilical hose. Thousands of litres of air could be compressed under massive pressure and held in a metal cylinder small enough to be carried on a diver's back. A valve mounted on these tanks reduced the pressure of the cylinder supply to something that could be breathed safely. Without this, breathing from them would be like inflating party balloons from the air hose at a petrol station – the uncontrolled rush of high pressure gas would soon injure the lungs.

As well as reducing the pressure, valves were also used to allow air to be supplied on demand – sipped from the tank like a drink from a bottle. This self-contained underwater breathing apparatus became known as SCUBA and allowed explorers and workers to greatly extend their freedom in the underwater environment. But having solved the problem of how to maintain oxygen supply, divers encountered a new threat. Staying safe at depth is about more than the availability of fresh air. Maintaining your supply of oxygen is only the first challenge. With depth comes pressure and with pressure physiological alteration.

~

Above water, you are swimming at the bottom of a different ocean: an ocean of air. You're not aware of its weight above you but it's there nevertheless, pushing on your body: 14.7 pounds for every square inch. We call that one atmosphere – appropriately so, because it is the pressure exerted by the weight of the single atmosphere that stands above us.

Water is far denser than air. The kilometres of gas above exert a pressure equivalent to just ten metres of sea water. Every ten metres that you sink into the water adds another whole atmosphere of pressure. So at ten metres the pressure is double that at sea level. At twenty metres it's three times as great, at thirty it's four times and so on.

Now water, of which the many trillions of cells of our body are largely composed, cannot be compressed and so for the most part our bodies do not significantly deform as we descend into the deep. But the same is not true of the pockets of air held in soft body cavities like the gut and the lungs.

Volumes of gas trapped in the body are reduced as pressure rises. If you were to blow up a balloon at the surface and managed to pull it down to ten metres it would halve in volume. And at twenty metres below the surface, where the pressure will have tripled, the balloon will have shrunk to a third of its original size.

And the reverse is also true. A balloon inflated under pressure at twenty metres will grow to three times that size by the time it has floated up to the surface. And if the skin of the balloon can't accommodate that increase in size then the balloon will burst.

This is among the most immediate threats that divers face. The lungs are effectively clusters of millions of airspaces – alveoli – that behave like tiny balloons. To inflate them at depth you need to take a breath of pressurised air sufficient to overcome the increased pressure of the surrounding water. Once inflated they'll behave exactly like balloons, expanding as you rise towards the surface. And with walls that are in parts a single cell thick they'll burst just as easily too.

A rupture of the lung's alveolar air sacs will push air into places that it doesn't belong. The real risk is of air expanding out of these burst balloons and entering the blood vessels that run over their surfaces. If this happens, the bubbles can travel through the veins of the lung back to the heart. From there they will be fired off around the body, blocking essential routes of blood supply. And if the circulation of the brain or the heart is involved, the effect can be instantly fatal.

Apart from drowning, this phenomenon of arterial gas embolism is the leading cause of death among divers. This is a consequence of what we call barotrauma – literally the wound of pressure. It is easily avoided. The trick, as with most things in life, is to keep breathing. Breathing in and out during the ascent releases the expanding gas and stops the lungs from over-inflating. But panic, followed by a bolt to the surface with your breath held, will kill you.

~

Pressure doesn't just change the volume of gas within our bodies. It changes the way gases affect our bodies. Air comprises 21% oxygen, 78% nitrogen and a mixture of trace gases. The nitrogen is usually inert, and at sea level pressure passes in and out of our lungs without any noticeable effect. But it doesn't remain so innocuous under pressure. Underwater, nitrogen acquires narcotic properties. The higher the pressure, the more intoxicating its effect becomes.

This, a phenomenon that divers call 'the narcs', feels almost exactly like being drunk and becomes very noticeable below depths of twenty or thirty metres. And while it might sound like fun, it's dangerous in a situation where small mistakes have the potential to be catastrophic. Oxygen too becomes less benign at depth. Under pressure it can become toxic – particularly to the lungs and the central nervous system – in the worst cases caus-ing seizures. And again the problems become more severe the deeper you dive. But some of the most severe medical problems associated with diving occur only after you've left the water.

Nitrogen is highly soluble. At sea level it passes from the lungs into the bloodstream until blood becomes saturated and no more can be taken on board. When we go diving the increase in pressure temporarily allows more nitrogen to dissolve, super-saturating the body and its tissues. This in part accounts for the narcotic effects seen at depth. But once the diver returns to the surface – and to normal pressure – this excess nitrogen must come out of solution and form gas. This is exactly the same process that we see in the fizzing of a bottle of cola when the cap is twisted off. Surfacing slowly is like releasing the pressure held in the bottle gradually and avoiding the overwhelming surge of bubbles.

We all fizz when we surface from a dive. The trick is to limit the number of bubbles and the rate at which they form. There are two ways of achieving this. Spending less time at depth reduces the amount of additional nitrogen that accumulates. Ascending slowly, or stopping on the way up, allows the shower of bubbles to be captured in the circulation of the lungs and filtered out of the body. This is the basis for the diving tables which define how long a diver may safely stay at depth and the rate at which they can ascend.

With proficiency and protocol comes an illusion of safety, but the dangers of subjecting your body to extremes of submergence are real and potentially deadly.

~

Christmas approached on Qalito. Weekends aside, we dived most days. The coral surveys saw us swim in teams of four, cataloguing the flora and fauna in locations that no one had ever dived before. The dives themselves were super-conservative; we were forbidden from hitting more than eighteen metres and we were never under for more than forty-five minutes. There was nothing particularly extreme about any of it. The water was warm and crystal clear.

It was a far cry from the dives I'd done in British coastal waters, where you sometimes had to wear a fleece jacket under your dry suit, were often unable to see your hand in front of your face, and at times had to grapple with ocean swell. Compared to that, the stuff we did on the expedition in the Pacific felt absurdly placid, little more than souped-up snorkelling. And yet the threat was always there.

In the run-up to Christmas we managed to squeeze in a couple of recreational dives. These were the highlight of the expedition – a chance to do some proper exploratory diving. The surveys were great, but being forced to follow compass bearings and record accurate fish and coral species counts on our slates somewhat detracted from the grandeur of the spectacle.

There was a clutch of newly discovered dive sites scattered around our island group that the expedition found and named. We were pretty literal about it. The reef in front of the expedition house was called 'House', another which was carpeted with sea grass became known as 'Garden'. The site we would dive that day was simply called 'Magic'.

~

The dive started well enough. We came off the boat as a group of four, split into two buddy pairs and began moving over the shelf, dropping close to the sandy bottom, drifting slowly along a line of sea fans.

I remember breaking into a broad smile as we settled into position. The reef swarmed with life. It was exploding with colour, as beautiful as anything I'd ever seen. 'Magic' indeed. There was a gentle current that pushed through alongside the coral shelf, bringing energy into the system: nutrients and foodstuffs, like a perpetual all-you-can-eat buffet. And on this the residents of this coral neighbourhood feasted. Everything was well fed, from the clownfish and anthia to the barracuda and sharks.

We floated on, throwing emphatic 'OK' signs to each other. There was no need to swim; the current was doing all the work. I sat, neutrally buoyant, taking it all in. No survey counts to perform, no compass work, just us and the ocean, flying along. The surface rippling above, with the sunlight playing through it, appeared deceptively close.

We passed by a chunky looking reef shark, its powerful, diamond-shaped body framed perfectly against the blue. There was no need to worry; it wasn't on my list of properly dangerous predators and besides, why would it bother taking us on when there was an endless supply of more predictable prey on the coral wall?

We drifted on. It was the sort of dive you fantasised about: a site virtually unexplored, unspoilt by tourists and queues of dive boats at the surface. The coral was lush and its neighbourhood in rude health. We were on our own here; this was *our* coral reef, unnamed and undiscovered until this expedition had happened upon it.

Then the current took us around a corner and the landscape began to change. I looked down. The reef was fading now, replaced by a sandy surface strewn with more barren-looking dead coral and rock. I was vaguely aware that we'd been picking up speed all along, but things really started moving after we made the turn.

I passed over one of the rocky outcroppings, too quickly for comfort, aware that my pace had suddenly accelerated. I turned to the rest of the dive group behind me and punched the open palm of my left hand with the fist of my right – trying to signal to them that I was worried about the current.

I needn't have bothered; they'd already worked that much out for themselves. All three were below me, hanging off the rock I had just passed, legs and fins fluttering in the current like flags in high wind. The current was growing steadily stronger, threatening to take us beyond the limits of the dive site.

I turned and tried to swim down towards them, descending again at a time when I should have been on my way back to the surface. A handful of metres separated us, and the current couldn't have been flowing at more than a few knots, but it felt like running into a gale. I inched forwards, trying to reach the relative safety of the rock. I was kicking furiously, my body out at full stretch, heaving great lungfuls of air from my tank, trying to make ground.

I was on the verge of being swept off by the flow and separated from the rest of the group. An earlier expedition had nearly lost a diver this way, pulled away from his buddies by the current. He'd been found many hours later, alone on the surface, more than a mile away, just before nightfall. They were lucky. When they got to him he'd all but given up hope of rescue.

I was keen to avoid the same fate. But here on this underwater treadmill, kicking as hard as I could, other risks were beginning to creep in. My air supply was falling faster than ever – I had to keep an eye on that. But at the back of my mind, I was aware that all of this frantic thrashing was silently accelerating the changes in my physiology caused by diving, increasing my risk of decompression illness.

~

When you exercise, your rate and depth of breathing go up. At the same time, the blood flow to your muscles and the capillary network around your lungs increases; your heart pumps harder and faster – chucking out larger volumes of blood – trying to keep up with demand.

I'd gone from sipping around half a litre of air from my tank with each breath, to perhaps five times that volume. My heart rate was up too. The five litres of blood ordinarily ejected by my heart every minute had now increased to four or five times that number.

All of this was an effort to step up and meet the demands of this new exertion: helping me to inch closer to that rock. What had started as a sedentary drift had now become a ferocious sprint. But in boosting the output of my heart, and the volumes of gas being shifted in and out of my lungs, I was – every second – bringing larger quantities of air into contact with a much greater flow of blood, further saturating my body with nitrogen. The simple fight or flight response that I needed to get myself out of immediate danger was adding fizz to the tissues of my body, trading improved performance in this moment for problems that I'd have to deal with later.

I reached them, finally, still breathing uncontrollably hard, paying back the oxygen debt. I scrambled to look at my air gauge. I had a little less than a quarter of a tank remaining.

We still had to work out what to do. Bryn, the most experienced diver among the four of us, scribbled some words on a slate while we held onto him. 'Near Wilkes Passage!!!' he wrote. I shrugged at him, unsure of what that meant. He scribbled some more. 'Shipping lane.'

We had no choice; we had to stick together and get to the surface and hope that we hadn't drifted too far from our boat. We aborted the dive, ascending into the pounding flow, sticking close to one another and stopping the ascent as we reached five metres' depth, hoping that this would be enough to help release the nitrogen that we'd stored up in our bodies.

I imagined the frothing of my blood. The nitrogen would be rushing out of solution now, possibly overwhelming the capillaries of my lungs. Usually the three minute stop at five metres was just an added precaution, something you did to make sure that you were absolutely covered. But I wondered this time if it would be enough. We stayed as long as we could at five metres.

We had the luxury of no more than a few extra minutes; my air was nearly finished and we were still being carried by the current.

I had time enough to think about where we were. Our shallow excursion of a few metres into warm, clear water suddenly looked far less benign. At the surface we'd have to hope that the dive boat could find us. It was already late afternoon and the Sun would set fast.

And then there was the decompression. Even if I felt OK it would probably be more than twenty-four hours before I knew whether I'd managed to deal with the additional burden of nitrogen. The symptoms of decompression illness – the numbness, the tingling, the pains in your chest, the difficulty with breathing, the creeping disabilities – all of that could evolve at pretty much any time while the fizzing in my bloodstream continued.

Nothing in my knowledge of dive medicine was particularly reassuring. We had stuck well within the limits of the dive tables, but more than half of all episodes of decompression illness happen to people diving within recommended limits. And I had just compounded the issue by exercising for a couple of minutes at top whack, loading my tissues and blood with far more nitrogen than I would have on any ordinary dive.

The nagging uncertainties that come with diving are easier to dismiss when you're within close reach of a recompression chamber. But out here it would take the best part of a day to get to the mainland and into a place with that kind of kit. And that journey could only safely happen during daylight.

There was no fall-back position here, nowhere to go if I really was sick. My body had started changing the instant I'd entered the water, adapting to the new environment, yielding to the pressure and the changes that this inflicted. It suddenly seemed ridiculous to have carted a bag full of diving gear out into the middle of nowhere just to scratch the surface of the ocean.

The gadgetry that allowed me to descend a few metres beneath the water left me naked above it and far from help. When we broke the surface I had never before been so happy to spit my regulator out and breathe ordinary air.

After a mercifully brief search, the dive boat found us. We drove back to Qalito, arriving just before sunset, happy – for once – to be out of the water. I spent the evening in my bed on the beach, reading my dive medicine textbook by the light of a head torch. Having pushed the most sublime edge of the envelope that supports human life – using a twenty-first-century sports kit I'd bought in a dive shop in London – I was left underneath that palm tree in darkness hoping that the envelope didn't push back, little better equipped than a physician from the 1900s. I was hopeful that this time I'd get away with it, but in that moment it seemed like tremendous folly.

∼

Human life is supported in only the narrowest margin around our planet; the water that covers fully four-fifths of its surface doesn't support the immersed explorer. And I'm not just talking about the absence of breathable oxygen. Taking air with us into the deep doesn't allow us to extend our stay indefinitely. We change as we dive, adapting to the new environment, but the water column also changes us, as do the gases that we take into our lungs. We're not supposed to stray from sea level. At least that's what our physiology tries to tell us. We're hopeless in the water without support – not much better even with our own oxygen. We should stay on dry land, somewhere warm. We are tropical animals after all.

Above the surface of the ocean, helicopter platforms open up avenues of exploration and rescue unthinkable in earlier centuries, and with an immediacy unparalleled in any other age. Borne at high speed out over the tropical seas of the Pacific or the icy waters of the North Sea, we rely upon that remarkable engineering to extend and protect us. It takes mere minutes to segue from the warmth and security of an airport terminal to flight above an environment entirely inhospitable to human life. And all that lies between you and miles of open water is an aluminium can suspended beneath an absurd engine that throws air at the ground in an effort to defy gravity

With helicopters and cylinders of air we have projected ourselves over and beneath the ocean; extending our capabilities in rescue and exploration. But those innovations provide only the barest protection from environmental extremes. Mishaps are promptly punished and rarely forgiven. Augmented by technology and engineering we simultaneously become more capable and more vulnerable.

This is the lesson I learn from this apparently benign expedition. Even here on this paradise island we are horribly exposed, remote from the artificial carapace of medicine and science that we've constructed to try and protect us from our mistakes.

∼

Qalito grew steadily hotter as the expedition wore on. Despite the odd decompression illness scare, I became less and less worried about the remoteness of the survey team. We were better served than most explorers, with creature comforts that at times made it feel less like an expedition and more like a five-star retreat. And yes there was risk but this was exploration and the two couldn't be divorced.

Most of the volunteers departed for cooler climes in late December, leaving a die-hard core behind. On Christmas Day we broke expedition rules and pulled a couple of barracuda out of the water on a line. It wasn't turkey but then we didn't have gravy, sprouts or carrots either and it beat the hell out of the staple diet of packet noodles and tinned tuna we'd eaten every day since we arrived.

We continued to dive each day, finding new reefs to catalogue. The oceans were changing: warming and slowly acidifying – processes that stunt the growth of coral and the ecosystems swarming around them. It wasn't a process you could see with the naked eye. The expedition and all its data provided but a snapshot, a single frame in the decades-long, time-lapse photography that charted the effects of global warming on this single aspect of marine life.

On New Year's Eve we gathered on the beach close to midnight and stared at the stars. It was perfectly dark; the band of the Milky Way ran clear across the sky, as spectacular as I had ever seen it.

We sang 'Auld Lang Syne' and shouted 'Happy New Year!' into the darkness. Occasionally we'd stop and listen, thinking we could hear voices in the distance. There was no light to be seen out to sea. But we knew that we weren't alone, that somewhere across the water there were other islands like ours out there. So we continued to shout, never sure if we were loud enough to be heard or if we had made ourselves understood, never sure if we were listening carefully enough. And above us lay another ocean of impossible vastness, about which we had precisely the same uncertainty.

CHAPTER 7
ORBIT

31 January 1977: horse and rider watch as the space shuttle Enterprise is towed through California from the Rockwell construction facility site to Edwards Air Force Base for a year of flight tests

WHEN I DECIDED that after my astrophysics degree I'd turn around, head back through the revolving doors of my university and sign on to study medicine, I had to break the news to my parents. They cried. Like many immigrant parents, I suppose, they'd always dreamed of their son becoming a doctor. My bank manager was close to tears too, though for different reasons.

By the third year I was running out of cash. I spent weekends filing slides in a photographic agency, worked shifts as a doorman at the student union (it was Universiy College London – you hardly had to be Arnold Schwarzenegger to pull that role off) and even had a laughably short and shockingly bad stint as a DJ. None of it quite paid the bills.

But I had a plan. NASA, I reasoned, was a multibillion dollar agency in the business of launching human beings into space. And if they were doing that then they must need doctors. And if they needed doctors then surely, I thought, they must have pots of cash to fund people like me.

The mismatch between my expectations and brute reality was cosmic in scale. NASA, of course, had billions of dollars but they were all spent – and then some. They had no bursaries either and, even if they had done, my British passport wasn't going to help me get to them. NASA as a federal agency of the United States is forbidden, under executive order of the President, from employing non-Americans. Most of the replies to my enquiries made that point none too subtly.

I gave up on the idea of getting a grant and embraced the Armageddon of growing student debt. I decided instead to send dozens of letters asking for the opportunity to spend some time as an intern with a NASA research lab. There was no money in this – I'd have to do it for free – but at least I'd get up close to the place that had fascinated me throughout my childhood.

Here too I was met with a barricade of polite refusals. Then came the age of the dial-up modem and I started blizzarding out emails. I was lucky. It was a time before people had figured out spam filters. Somewhere, somehow, one of those letters, faxes or emails got me an application form – and, unbelievably, that application form won me a place on an aerospace medicine course at Johnson Space Center in Houston.

~

Johnson Space Center was like Disneyland for adults, only the rides were better. We spent mornings being briefed by astronauts, flight surgeons and engineers, and afternoons touring the facility. It was 1997 and NASA was preparing to begin assembly of the international space station. The centre itself was strangely underwhelming in appearance, resembling a 1970s university campus, with an uninspiring brown-on-beige-on-grey colour palette. The scattered low-rise buildings were largely windowless, and distinguished from one another only by numbers. Driving along the network of roads inside the gates you could be forgiven for mistaking it for just another American Corporation of Making Everything industrial complex.

But inside those buildings the stuff of space flight happened. People talked seriously about how they might best get to Mars or back to the Moon; engineers built prototype plasma drives and novel hydroponic life-support systems. It was an industry, it appeared, that encouraged its employees to have crazy ideas and then make them happen.

The medical operations group did a good line in humility. One of the first things they explained was that at NASA doctoring came a poor second place to the business of engineering. And even this lowly position was only recently attained. Accounts of the early days of human space flight show that aerospace medicine used to be barely tolerated. The docs were seen as an obstacle to the job of flying, neither necessary nor useful. Joe Kerwin, a former Skylab astronaut and NASA's first physician in space, summed it up neatly: 'You gotta understand, the crews won't be happy until the last psychologist has been strangled on the entrails of the last flight surgeon.'

Happily, relations between the astronaut corps and the flight medicine clinic improved steadily as the years progressed. But medicine would always play second fiddle to NASA's all-dominant culture of engineering, even though the challenge to the human body presented by the environment of space couldn't be more extreme.

～

If you rise up through the atmosphere from sea level, the going gets tough long before you get anywhere near space. Anything above 5,000 feet counts as 'high altitude' as far as physiologists are concerned. Even at these modest heights the medical problems caused by altitude can begin to develop.

Once you get to around 29,000 feet, just five and a half miles above the ground, you reach the highest point on the surface of the planet: the summit of Mount Everest. This appears to be very nearly the high-altitude limit for unsupported human life. A couple of hundred feet higher and the mountain would be unscaleable without supplemental oxygen.

Mountaineers arriving at the summit of Everest do so only barely alive, having altered their physiology over weeks adapting to the challenges presented by the rarefied atmosphere. Here, with or without oxygen, every step becomes a task of Herculean scale. Summiteers describe the excruciatingly slow plod along the last ridge that stands between them and their goal each stride punctuated by great gasping bursts of hyperventilation as they struggle to repay the oxygen debt incurred. Even after weeks of adaptation, their bodies are only just capable of this feat. An unadapted individual, who hadn't endured the weeks of acclimatisation, would be incapacitated in less than thirty seconds by exposure to the same altitude.

A typical commercial jet airliner cruises at around 36,000 feet – a few thousand feet higher than the summit of Everest – but the passengers and crew within are breathing normal, low-altitude air. It is only pressurisation of the cabin that leaves them able to enjoy in-flight movies and moan about the lack of leg room, rather than loll around unconscious in their seats as a prelude to death from hypoxia.

For it is reduction of pressure that causes us problems at high altitude. With fewer molecules of oxygen in every breath, the pressure exerted by the oxygen in our lungs falls and so too does the rate at which it passes across the membranes of the alveoli and into our bloodstream. This leaves our red blood cells, and therefore our tissues, starved of oxygen. You can compensate for that in one of two ways: either by pressurising your environment – as commercial airlines do – or by increasing the amount of oxygen in the air that you breathe.

Commercial airlines rely upon pressure to keep their passengers properly oxygenated. In pre-flight safety videos flight attendants calmly show off the yellow oxygen mask that would pop out of the ceiling and dangle above your seat should cabin pressurisation fail. Part of their briefing urges you to behave selfishly, asking that you put your own mask on before attending to anyone else. But there's a good reason for this. At 36,000ft, in the absence of supplementary oxygen, a sudden loss of cabin pressure will incapacitate you in less than thirty seconds – roughly the time it would take you to fight a recalcitrant toddler – by which time both of you would be left helpless.

Things only get worse as you ascend. Pilots of unpressurised aircraft have to compensate for the reduction in atmospheric pressure as they climb higher by increasing the concentration of oxygen that they breathe. The lives of the Second World War bomber crews, flying at altitudes of up to 40,000ft, depended as much on the oxygen supplied to their face masks as they did on avoiding flak batteries and enemy fighter cover.

In unpressurised vehicles the higher you go, the greater the concentration of oxygen you require in the gases that you breathe. But above 40,000ft even pure oxygen isn't enough to keep you alive. At this altitude the pressure falls to less than a fifth that at sea level. Here the oxygen doesn't exert enough pressure to drive itself across the membranes of your alveoli and load the molecules of haemoglobin in your bloodstream.

To support human life at these higher altitudes, oxygen must be breathed under pressure. These more advanced oxygen systems – the type used in modern jet fighter aircraft – comprise masks which form an airtight seal around the face and then force oxygen into your lungs at a huge rate of flow. Wearing one feels like sticking your head out of the window of a car thundering down the motorway and trying to breathe against the rush of air. The effect is to inflate your lungs like a balloon, raising the pressure within them above the ambient pressure of the air outside; facilitating the loading of haemoglobin with oxygen and thereby ensuring your survival. And even this only works up until a point.

Above 63,000 feet you encounter the Armstrong line, an

atmospheric limit above which the poor oxygenation of your bloodstream is no longer the only factor threatening your life. (Although the Armstrong limit refers to a space-flight boundary, it takes its name from aviation physiologist Harry George Armstrong, as opposed to he of the 'one small step'.)

The Armstrong limit is essentially the altitude at which you begin to boil yourself. Let me explain that statement a little. Pressure cookers work because the boiling point of water, and all other liquids, rises as ambient pressure rises. Your carrots cook more quickly in a sealed cooker because the pressurised water inside is able to reach a temperature higher than 100°C before it bubbles and boils. The reverse is also true: the boiling point of liquids reduces as the pressure falls.

At the summit of Everest water would boil at a little over 70°C. At around 63,000, feet the boiling point of water falls further, to 37°C: the same as the human body's normal core temperature. At this, the Armstrong limit, water contained in the tissues of the body spontaneously begins to boil. Bubbles of vapour evolve and expand, swelling soft tissues, causing the body to balloon. Interestingly – and contrary to sci-fi lore – the blood in your arteries doesn't boil. The muscular walls of those vessels behave like a crude pressure cooker preventing the water in the arterial bloodstream from boiling.

But in the veins the story is different. Here the blood flows at much lower pressures and bubbles of water vapour can and do form. With longer exposure to high vacuums, these bubbles grow and cause airlocks, bringing the circulation to a halt and eventually causing cardiac arrest. To avoid this fate, people venturing above the Armstrong line must swap their oxygen masks for pressure suits, surrounding themselves entirely with an artificially created sphere of survival. This is part of what we come to expect of spacemen: astronauts in helmets and bulky sealed suits, insulated against the ravages of space, taking a little bubble of the Earth's atmosphere with them.

The Armstrong limit defines the height above which simple augmentation of physiology is no longer enough. Beyond this human life depends entirely upon artificial life support for survival. And that layer around the Earth, just twelve miles high, represents the narrowest of slivers. If the Earth were the size of a football, then the zone in which life exists unsupported would be little more than a film less substantial than a sheet of paper wrapped around its surface.

Space begins at an indefinite point. For physiologists it is the Armstrong line limit that marks its threshold, but for aircraft engineers it starts at the Von Karman line one hundred kilometres above sea level. Here the atmosphere is so thin that ordinary aircraft can no longer push against it to steer or generate lift. To the physicist, true space starts many thousands of miles away where the statistical probability of collision between two gas molecules becomes insignificant. But for astronauts it's not about altitudes or pressures. For them the frontier of space, and all of its attendant risk, begins on the launch pad, from the moment the rocket engines light.

∽

I arrived in Florida at the beginning of July 2011, a few days before the big launch. *Atlantis* stood ready on the pad, waiting to carry its crew of four astronauts into orbit. She was the last of her kind: her sisters *Challenger* and *Columbia* had been lost to tragic accidents. *Discovery* and *Endeavour* had already been withdrawn from service and now lay stripped down in hangars being made ready for transport, preparing to take their place as historical exhibits in other cities. This mission was to be the last of the space shuttle programmes. After three decades and 135 flights, NASA had called a halt to the project.

On the morning of launch, the air outside was humid – heavy with the threat of rain. Tropical storm fronts had blown ashore one after another in the past couple of days, throwing lightning at the ground and drenching the soil. The weather around Cape Canaveral was always unpredictable in the summer: blue skies could turn to thundercloud grey in minutes, carrying sudden torrents of rain with them.

For the past twenty-four hours I'd been glued to meteorological websites, trying to make sense of isobars and radar pictures, watching fronts evolve out at sea and migrate inland. I wouldn't usually care, but today at 11.21 a.m. there had to be near cloudless skies above Kennedy Space Center for ten minutes. Whatever happened before or after that didn't much matter.

Within those ten minutes lay the launch window for *Atlantis*. They marked the fleeting period when the Earth would rotate Pad 39A into just the right position, so that when *Atlantis'* engines were lit the thrust would carry the spacecraft – and her crew of four – into orbit, to arrive at precisely the right place and time to allow her to rendezvous with the International Space Station.

The space station itself was travelling around the Earth at over 17,000 miles an hour. That huge velocity gave it enough energy to remain in stable orbit, allowing it to resist the forces that would otherwise bring it crashing back to our planet.

To catch up with that platform, *Atlantis* herself had to become a missile, acquiring enough energy to accelerate to the same speed. She would get a little kick from the Earth, borrowing some of the energy of its rotation. Like everything else on the surface of the planet, the launch site wasn't stationary. It was rotating with the Earth at a little over nine hundred miles an hour from West to East.

The rockets could make use of that, like a long-jumper starting their run-up on a supersonic conveyor belt. And while that sounds like a good start, the main thrust of the acceleration – the acceleration that would drive *Atlantis* to more than 17,000 miles an hour – had to be achieved through the brute force of rocket engines.

The environment of space is uniquely hostile. But when it comes to orbital space flight, the dominant threat to human life comes from the vehicles and their launchers and the way they behave. Two hundred and fifty miles, roughly the distance from the surface of the Earth to the altitude of the space station, doesn't sound like a long way. But rocket science isn't about distance, it's about defeating the force of gravity and the energy released in accomplishing that feat.

$\sim$

Atlantis was already standing exposed on the launch pad, towering over two hundred feet above sea level. Its fat, orange external tank had been filled overnight with hundreds of thousands of litres of liquid oxygen and hydrogen. Those cryogenically stored fuels, sealed in the insulated tank strapped to *Atlantis'* belly, were gently boiling off.

At the pad, the stack was creaking and groaning, straining with the competing thermal stresses of the freezing fuel and muggy warmth of the Florida air. Elsewhere hoses hissed and vapours poured forth. At launch that liquid fuel would feed the three shuttle main engines which sat in a cluster at *Atlantis'* rear.

Flanking the tank and the orbiter were the two solid rocket boosters (SRBs). Nearly four metres across and about as long as an Olympic swimming pool, those cylinders were filled with five hundred metric tonnes of ammonium perchlorate blended with aluminium: an explosive combination studded with oxygen atoms whose energy was just waiting to be released.

That material was combined with a binding agent leaving it in a solid state, with the consistency of putty. When lit it would burn at temperatures comparable to those of the surface of the Sun and massively augment thrust in the first two minutes after ignition.

Atlantis had stood waiting on the pad for several weeks, undergoing meticulous final preparations. The orbiters returned from space nearly dead: gliding without power, bodies scorched, fuel and energy spent, engines thrashed to the limits of their endurance. For the hundreds of engineers responsible for turning them around again and returning them to flight, it was an act akin to resurrection.

Tonight was the first time in this mission that *Atlantis* had been fuelled ready for launch. They had evacuated the perimeter as far back as two miles to all but the most essential staff. *Atlantis* was dangerous now; the potential energy stored in the chemicals of her external tank and the solid fuel of the rocket boosters were enough to propel the two thousand tonnes of stack into space at twenty-five times the speed of sound. The groaning, the creaking and the hissing may have been caused by expanding gases and grating metal, but even the more seasoned engineers regarded *Atlantis* as though it were an animal slowly coming to life, with a personality of its own.

On the morning of launch, contrary to all expectations, the skies miraculously cleared. We stood with our hearts in our mouths as the digits on the clock ran down. A last minute debacle on the launch pad very nearly led to the launch being scrubbed. But *Atlantis* was ready and nothing would now stop her.

The countdown clock ticked down to zero. We stood and watched as *Atlantis* rose into the sky. It felt wrong; launches always did. It was an event on a scale that didn't otherwise exist in the world. A massive object racing straight up, far faster than it should be able to, burning engines bright enough to light the entire bank of clouds into which it eventually flew, disappearing in seconds. I stood, breath bated, until the solid rockets separated.

The crew on board knew the risks of their endeavour better than most. As you climb on the highest slopes of Everest there are points at which you pass the bodies of people who have died on the mountain – a sobering reminder of the consequences of taking such risks. In a similar way, astronauts riding aloft are aware that as they hear the words 'go at throttle up' they are passing the point at which *Challenger* failed and they know, as they decelerate through Mach 19 on re-entry that *Columbia* got here – and no further.

~

The trick to flying is to throw yourself at the ground and miss. At least that's how Douglas Adams explained it in the *Hitchhiker's Guide to the Galaxy*. And while his description was constructed for comic effect, it actually captures – in a strangely accurate way – what astronauts heading into orbit actually do. They climb into vehicles, fire their rocket engines and hurl themselves across the Earth until they run out of planet to fall onto. And once there they continue to fall freely around the globe, held by the bond of gravity, unable to escape the Earth's grip or return to its surface. The term 'orbit' simply describes the act of falling towards a celestial body without ever hitting it.

There is of course a little more to it than that. The art of rocket science is a discipline filled with everyone's worst maths-class nightmares: calculus stacked upon the mechanics of circular motion framed within exotic co-ordinate systems. And when you get down to the nitty-gritty of building the things, there's a load of pretty nasty chemistry to bend your mind around too.

The reality is worse still. The stuff on paper has to be engineered to work in the real world without all of the simplifying assumptions. The nuts and bolts have to travel at many thousands of miles an hour and then fall gracefully through space, precisely as predicted, without flaw or failure. The only way you could make rocket science any more daunting as a prospect would be to add humans into the equation as passengers.

This then is the challenge of human space flight. No amount of adaptation or acclimatisation can prepare the body for exposure to hard vacuum. No amount of augmentation of physiology can make that environment survivable. Instead, bubbles of life support must be artificially created, maintained and sealed against the exterior. And these must then be crammed intact into the architecture of a space vehicle small enough and light enough to respect the great energies demanded by orbital space flight but spacious enough to afford at least rudimentary comfort for the crew. When it comes to human space flight throwing yourself at the ground and missing is only half the battle.

❧

As a junior doctor I'd occasionally get a chance to spend some time at the Cape, working and researching with the medical team there. Formally it was Kennedy Space Center, NASA's space port, the point on Earth from which every human-rated American space vehicle had ever departed. But to me it was always the Cape. It sat a few dozen miles outside Orlando on the eastern seaboard of the United States, a sprawling government complex reclaimed from wet marshlands in the 1960s for the purpose of doing something outrageous with explosive rocket technology.

As a facility it was vast: hundreds of square miles of land that ran along the Florida coastline. From the rear gate I could drive at fifty miles an hour for the best part of fifteen minutes and still not reach the building in which I was based.

I'd drive up from my apartment on Cocoa Beach, be waved through the gate and get onto the causeway that spanned the Banana River. There I'd roll the windows down and crank the radio up. For a couple of miles I'd pelt across the water on that narrow stretch of road, with launch pads to the north and rocket assembly buildings in the distance, sitting between me and the Atlantic Ocean. It was the best part of every day.

From time to time NASA ran training courses for the civilian medical teams who might be called upon to attend a shuttle accident. We'd gather in lecture theatres and receive instruction on the anatomy and physiology of the space shuttle: how it might fail; what, in theory, we might do to help.

They showed us how the crew could escape a debacle on the launch pad by sliding down a 200ft-high zip-wire, getting from the crew deck to the ground in a few short seconds, crashing into a net and then bailing into an armoured car that they'd been trained to operate. In an emergency they were told to climb in, drive straight through the perimeter fence and keep going in the hope that they might outrun the fireball and blast that would accompany the simultaneous detonation of a few hundred thousand litres of rocket fuel.

They showed us too that the shuttle could abort after take-off during its ascent. Redundancy was the name of the game here. After a few minutes of flight the mission could tolerate the failure of one of the three shuttle main engines and still get into space, albeit at a lower than intended orbit.

Losing an engine early, before momentum had had time to build, or losing more than one engine, would be a different matter. Unable to develop the altitude or velocity required to achieve low Earth orbit the shuttle could perform a Transatlantic Abort, a manoeuvre that would see it ditch its external tank and solid rocket boosters and vault across the Atlantic ocean, landing somewhere in Europe.

That journey across the Atlantic ocean of more than four thousand miles, which a commercial airliner would take perhaps eight hours to cover, would be completed by an aborting shuttle in less than thirty minutes.

There was another, even more outlandish scenario called the Return To Landing Site. Here, having lost an engine early in the launch, unable to make it to space but still strapped to its external tank and two solid rockets, the shuttle could – in theory – attempt to flip itself over and fly back to Kennedy Space Center. During this abort the shuttle would jettison the solid rocket boosters after two minutes. Then, still strapped to the external tank and at this point heading towards Europe at several thousand miles an hour, it would ascend and use its manoeuvring thrusters to flip itself over, rotating through 180 degrees like a pancake, with its remaining main engines still burning.

Having performed the equivalent of a supersonic handbrake turn, the shuttle's momentum would continue to carry it on towards Europe.

Flying backwards, with its nose pointing roughly towards the

United States, the engines would be facing the direction of travel, thus slowing the shuttle down. At some point the shuttle – rocket motors still firing and external tank still attached – would see its progress towards Europe arrested. Momentarily it would come to a standstill before accelerating once again, this time back towards the States. The crew would then dump their external tank and attempt to glide unpowered back to the site from which they'd launched some twenty-five minutes earlier.

It wasn't just failure of the engines that could lead to these emergency aborts. Both the Transatlantic Abort (TA) and the Return To Landing Site (RTLS) could also be used to get the shuttle back on the ground quickly if a significant failure in the life-support system occurred. There was after all no point in parking a vehicle in perfect orbit if the crew inside could not be kept alive.

A peppering of euphemisms accompanied these briefings. There was an anticipation that under such conditions both the vehicle and its crew might return in 'sub-optimal condition', that the landings might be 'off-nominal' in character. Behind this technical phraseology lay the risk that during an abort the shuttle might crash on or short of the runway and the crew might be severely injured in the process.

To civilian clinicians these abort modes sounded like the stuff of science fiction. Even among the astronaut corps there was a little scepticism about just how successful a real RTLS abort might be. Nevertheless they dutifully drilled and trained for the scenarios, sitting fully suited in simulators for hours at a time, rehearsing their worst nightmares.

I often wondered why they bothered to do this, when the risk of these types of failures was so low and the chances of recovering intact after one of the more elaborate aborts was lower still. But like so many other things in exploration and medicine, they did it because the only other alternative would have been to do nothing – which for them wasn't an option at all.

∽

Even if the launch goes smoothly there is still the possibility that a medical emergency might arise during a mission, far from the safety of any hospital. Because of this, considerable effort has been invested in designing avenues of escape and medical contingencies for space crews. People have even gone so far as to devise ways of resuscitating victims of cardiac arrest. This is no mean feat. Imagine, for a moment, trying to deliver cardiac compressions while floating weightlessly in orbit.

The trick, it turns out, is to strap the patient to the floor of the vehicle, put your hands on their chest, brace your feet on the ceiling, and then use your legs to provide the necessary force. This has been trialled on resuscitation dummies in weightless training aircraft and it works surprisingly well. But if you're going to plan for the possibility of cardiac arrest, then you've got to consider precisely what you're going to do after the patient's heart starts beating again. Contrary to what Hollywood would have you believe, people who survive an arrest of their heart very rarely sit up the instant their pulse returns as though nothing had happened. The experience of total circulatory arrest, along with whatever it was that stopped their heart in the first place, tends to leave them critically unwell. Afterwards a period of extreme instability and a lengthy stay in an intensive-care unit is the norm. For all its sophistry, the International Space Station has less medical equipment and expertise than the average London ambulance. Definitive medical care is only available back on Earth.

It was predicted that during its lifetime of operation there would be at least one major medical incident aboard the International Space Station that would require evacuation to Earth. To cater for this, NASA started work on a new experimental vehicle: the X-38.

∽

There is a black and white picture from 1977 of the prototype space shuttle *Enterprise* being carried along a Californian desert road on the back of a huge articulated truck, being delivered to NASA's Dryden Research Centre for flight testing. Behind it is a snaking line of 1970s motor vehicles. In the foreground sits a man astride his horse, its breath misting in the cold January air. It is a picture of the old world watching the future arrive. And that is what we came to expect from the Space Agency. That's what NASA did: it served up the stuff of science fiction on the back of a flatbed truck and told you that *this* was what the future was going to look like.

I was reminded of that image when someone showed me the plans for NASA's new X-38, back in 2001. It was a wingless vehicle, shaped like a shuttlecock split in half through its nose; a wedge too narrow to allow an adult to stand upright inside, about the size of a luxury speedboat. It was windowless and profoundly alien in appearance. I remember thinking that if it landed unannounced in your back garden, you'd be pretty disappointed if something didn't then slither out and say, 'Take me to your leader.'

The X-38 was destined to be NASA's Assured Crew Return Vehicle – a way of solving the problem of what to do if an astronaut crew had a really bad day in space. The plan was to load it into the payload bay of a space shuttle, deliver it to the space station and then leave it docked until called upon.

In the event of some catastrophic failure of systems aboard the space station the X-38 would become a space lifeboat. The crew would scramble inside, lie down, strap in and punch out. It was a remarkable design, able to accommodate a crew of seven, shaped so that it could be steered in the upper atmosphere while travelling at hypersonic speeds and then endowed with the world's largest para-foil – a steerable canopy that would slow its descent to the ground to ensure a gentle landing. But it was intended to be more than just a fast ride home. In a medical emergency, with members of the crew critically ill or injured, it would essentially perform as a space ambulance, capable of being kitted out with medical oxygen, state-of-the-art patient monitoring and even ventilators.

But as costs mounted and the International Space Station ran into financial trouble, NASA was forced to make cuts. The X-38 was shelved and NASA returned to relying upon the *Soyuz* space capsule as their means of escape. Much smaller than the

X-38 and capable only of accommodating three crew members at a time, it was a lifeboat with no real medical capability. But in low Earth orbit it had become increasingly clear that the dominant threat to human life would not come from crew injury or malfunctioning physiology. There was something that doctors and mission controllers on the ground feared far more than any medical emergency: a catastrophic failure of the vehicles that carried and protected their astronaut crews.

~

Soyeon didn't plan on being an astronaut; she lived in South Korea, a country with no human space exploration programme. She watched science-fiction films as a child and fantasised idly about the possibilities of space, but her ambition went no further than that.

She was in her final year of PhD study when the adverts first appeared in newspapers. South Korea was to run a national competition, casting the net wide in search of the country's first astronaut. The contest had all the trappings of an *X-Factor* game show: eliminations would be run week after week, over four months, and the competition would be televised. To take part, the only prerequisite was that you had to be over nineteen years old.

Soyeon decided to apply, knowing that she couldn't possibly be successful. She was a twenty-eight-year-old laboratory scientist working on a graduate degree in bioengineering at the prestigious Korea Advanced Institute of Science and Technology

(KAIST) but didn't kid herself that she was anything special. She filled in the form anyway. It would be an experience just to be in the running, and a welcome distraction from the final year of PhD study. By the time the closing date for entries arrived in September 2006, 36,000 South Koreans had applied.

~

The mountain of application forms was screened – excluding those without the right educational background or qualifications, driving the numbers down to something more manageable. A 3.5 kilometre run then served as another coarse filter, this time for standards of physical fitness. The list of hopefuls thinned out quickly. By the end of the first month of selection there were only 245 people left – Soyeon among them.

Medical examinations, psychological evaluations and interviews filled the month of October. When Soyeon made it down to the final thirty candidates, she allowed herself the faintest glow of hope.

In November and December came successive rounds of televised elimination. As the tests came and went, Soyeon found herself still in the running. The tasks became more elaborate. The contestants experienced weightlessness aboard a roller-coaster airline ride, dived in swimming pools to simulate space-walks and neutral buoyancy, and underwent decompression training. The superficial gloss – the studio lights, the spectacle and the telephone voting – was just that. Underlying all of this was a rigorous process of technical selection, of the type that any country might use to select professional astronauts. By the time the ten finalists lined up before the live television cameras on Christmas Day 2006, the assembled hopefuls looked much like the shortlist for any formal astronaut corps: a clutch of scientists, engineers and pilots.

There were to be two winning candidates, a man and a woman. Ko San, a thirty-year-old researcher at the Samsung Advanced Institute of Technology, was the successful male applicant. And standing next to him, blinking in the studio lights when her name was called, was Soyeon Yi.

Things moved quickly after that. Soyeon was told to halt work on her PhD and get ready to report to Star City in Moscow for training. The pace was bewildering. It was the end of December and they were due to report for training in Russia in three months' time. At that stage she didn't speak a word of Russian and hadn't yet finished her degree but none of that appeared to matter to the competition organisers. She was going to Moscow.

～

Soyeon's first memories of Moscow were that it was grey and bitingly cold. There in Star City, in parallel with an onerous training regime, Soyeon would finish her doctoral studies. She became a confident Russian linguist, endured survival training and got to grips with the culture of Russian cosmonaut training. There was initially, she felt, a dismissive attitude towards her from the predominantly male training staff in Russia. But Soyeon was thick-skinned and more than used to handling this sort of behaviour. Throughout her engineering studies in Korea she had pursued courses where women were in the minority and men were often less than progressive in their attitudes. To her Star City felt little different.

More attention was lavished upon Ko San. Although both Koreans were being trained, only one would eventually fly to the space station, and it appeared to be a foregone conclusion that it would be Ko San and not Soyeon.

After a year of training, when the time came for flight assignment, Soyeon's suspicions were confirmed. Ko San was awarded the prime slot. Soyeon was to be the backup crew member and, as such, never likely to fly in space. She had loved her experience nevertheless; it had transported and transformed her. Life, she felt, would never be the same again. Meanwhile Ko San prepared for launch, looking every bit the national hero that South Korea had sought to create.

And then came an unexpected turn of events. The Russian training teams are notoriously unforgiving of protocol violations. And though the details remain unclear, Ko San somehow managed to anger his Russian hosts. With three months to go before the mission, he was taken off the flight and in his stead Soyeon was promoted to the prime crew.

At first she was incredulous. She had never really expected to fly, yet now here she was, the prime candidate, due to launch in less than a hundred days. Usually adaptable, Soyeon was worried that she couldn't adequately prepare in that short time.

This fear continued to occupy her mind as the emphasis of her training changed focus and took on a new seriousness. And then, while having supper one day in Star City, she received a special phone call – one coming live from a module in space. It was Peggy Whitson, NASA astronaut and current space station commander, who was already in orbit. Peggy had heard about the last-minute change in the crew assignments and wanted to reassure Soyeon that she was good to go. During her training Soyeon had particularly looked up to Peggy. She noticed that wherever the American astronaut went, people appeared to respect her authority. That – Soyeon noted – was rare for anyone and rarer still for a female crew member. If Peggy thought that Soyeon was ready, then maybe she was.

~

On 8 April 2008, a little over eighteen months after Soyeon had first replied to an advert in a newspaper calling for astronaut hopefuls, she launched from Baikonaur Cosmodrome in Kazakhstan aboard *Soyuz TMA-12*. They took a handful of minutes to climb more than two hundred miles into space. Two days later, their capsule crept towards the International Space Station and docked.

Soyeon's time on the space station felt like a surreal dream. The assembled modules, joined end to end, gave the crew a free-floating space comparable to that of two commercial airliners. From the outside it appeared larger still. With its solar arrays unfolded, the station covered an area in the sky the size of two American football fields. Inside, the noise of its power and life support systems throbbing away was at times loud enough to make ear defenders necessary. It was a reminder that this was more than an assembly of buildings floating in orbit. It was a machine in which people lived – one that, through energy and ingenuity, created an artificial island of human survival in an otherwise uniquely hostile environment.

Soyeon busied herself performing a long list of experiments, taking time out during her ten-day stay to broadcast to school children and the wider South Korean public. She took the opportunity when she could to steal time in her cabin with its tiny window that looked out at the blue globe of Earth below. All too soon, it seemed, it was time to leave.

～

On the day of departure the crew crawled into the confines of the *Soyuz* capsule. They had to enter in strict order. Peggy Whitson entered first, cramming herself into the left seat. Soyeon followed, finding the rightmost chair. Finally Yuri Malchenko, who would command the *Soyuz* capsule on its flight back to Earth, wedged himself between the two. They completed their checklists, and the colleagues that they were about to leave behind as the new space station crew closed the hatches and sealed them in. There they sat in their bulky pressure suits, contained within their tiny bud of life support, suspended below the International Space Station.

The *Soyuz* backed off carefully from the station, creeping away at inches per second. There were no forward windows on the capsule; the crew's view through the small portholes was restricted and mostly looked out to the left and right side. From her seat Soyeon could see the vastness of the International Space Station (ISS) as it slowly receded. For all its artifice and fragility, the space station was an island of security compared with their tiny home-bound craft.

The trio hovered below the relative safety of the space station, separated from the ground below by a dense atmosphere and the need to bleed off the tremendous energies they had acquired at launch. They continued to pull away cautiously, taking nearly two and a half hours to put only twelve miles between themselves and the space station. This excruciatingly slow choreography underlined the vulnerability of both vehicle and station. The structure and the systems of both the *Soyuz* and ISS were finely balanced. Neither were designed for hard collision.

At a safe distance and on schedule, they fired the *Soyuz*'s rocket motors, slowing themselves down, giving gravity a chance to capture them more firmly. The *Soyuz* craft comprised three sections. At the front was the oval-shaped orbital module, accessible to the crew only while aloft. Behind it was a cone, the lower half of which housed the propulsion module. In the top part of that cone lay the re-entry module: a tiny, bell-shaped vehicle into which Soyeon, Yuri and Peggy were crammed. Superficially it resembled a giant pawn, taken from a chessboard the size of a football pitch.

Shortly before re-entry the crew capsule separated from the

other modules. Soyeon, sitting in the right seat, remembered her training for this phase of the flight. Specifically she recalled asking if she'd be able to see the orbital module as they separated from it. The answer was an emphatic *no*. Her instructor took her through the separation process again step by step, explaining that the modules would come apart like beads on a straight piece of wire. If she could see the module after separation it would mean that something had gone very wrong. And yet, after the pyrotechnic bolts had fired and the thrusters had begun to push them apart, she was sure she had caught a glimpse of part of that module through the porthole above her head.

Concerned, Soyeon reported this to Yuri. At first he thought that she must be mistaken. As the vehicle commander, he had been monitoring the instruments and all of them had registered a successful separation. He also knew that a near catastrophic failure in the separation process would have to have occurred for Soyeon to be able to see something of the separated orbital module from her seat position. Between them, Yuri and Peggy were among the most experienced astronauts in Russia and the United States. Soyeon, on the other hand, was a rookie and could have been mistaken. But then Peggy Whitson saw something through her porthole too, apparently drifting over and around their vehicle.

Strapped into their seats, with a limited view of the exterior, it was difficult to know what they had just witnessed. But whatever it was, they knew they shouldn't have seen it. Worse still, Soyeon now thought that she could see something flapping, still attached, against the outside of the capsule.

Re-entry started with the capsule 400,000 feet above the Earth. The weightlessness of orbital space flight was replaced by the forces of deceleration as the craft slowed against the atmosphere. Soyeon noticed that the ride was rougher than she'd expected it to be; the G-load seemed to be stacking on her chest faster and harder than the 4Gs she had anticipated. She reported this to Peggy, who tried to reassure her that the load was normal, and that the experience of ten days of weightlessness might make it feel more intense. But the G-load climbed quickly and soon even Whitson and Malchenko sensed that things were not right.

The three crew were crammed into the re-entry module, sharing just 3.5m³ of space – a couple of telephone booths' worth. They knew that the module's survival upon re-entry depended upon its ability to adopt exactly the right orientation – with its heat shield facing the direction of travel – as it passed through the atmosphere. It was not the physiological challenge of the space environment that threatened the crew here – it was the sheer violence of re-entry.

~

At launch, a vehicle like *Soyuz* must acquire enough kinetic energy to propel its crew at over 17,000 miles per hour. It does this exactly as a firework would, by liberating the chemical potential energy in the launcher and translating it into the kinetic energy of motion. The vehicle could in theory use the same process to slow itself down, but that would require another rocket motor of the same size that got it into orbit in the first place. To avoid having to carry that huge mass into space, the *Soyuz* slows down by losing energy to the atmosphere as it passes through it.

It's tempting to think that it is friction which slows the capsule's progress during re-entry. But that's not what happens. Instead, with the molecules of the atmosphere essentially unable to get out of the way as the re-entering vehicle screams through, a shock-wave of compressed gas builds up in front of the capsule. Much of the energy of motion is lost in heating that shockwave. The faster the capsule travels, the greater the heat generated. *Soyuz* is designed to stretch the re-entry out over a longer period of time, slowing down more gradually – a bit like the way a frisbee would sink towards the floor compared with a cricket ball. But even then the front of the capsule reaches temperatures of several thousand degrees – about as hot as the outer layers of the Sun.

As we learnt earlier in the book, human physiology functions very badly if the body's core temperature rises by just one or two degrees. People begin to die of heatstroke if it rises by more than three. The problem for designers of human-rated space vehicles is how to face a wall of heat of, say, 3000°C, and then park three astronauts behind it in a tiny capsule, maintaining that pocket and its system of life support at no more than 25°C.

This outlandish feat is achieved in two ways. First the base of the capsule, facing the shock front, is covered in a thermal shield. This layered surface sublimes, transforming from solid to gas as it heats, pushing the hot shockwave in front of the vehicle away as it does. The second element which allows the crew to survive the inferno is a precise angle of entry, which prevents the capsule from heating up too quickly and allows it to fly with the heat shield facing the direction of travel.

But Soyeon knew that they hadn't separated from their orbital module correctly. And that whatever it was that still remained attached could throw things off, leaving the capsule in the wrong orientation as re-entry began. If an unshielded part of the capsule was facing forward as they pushed through the dense atmosphere, the heat would very quickly destroy them and their vehicle. And if this had happened then the first indications would be a sudden increase in the G-load followed by heat building up inside the capsule.

Inside the capsule the G-meter, measuring the severity of their deceleration, peaked at 8.2G – more than twice the normal value – and Soyeon struggled to remain composed.

~

It is an old adage that the two hardest feats in all of rocket science are starting and stopping. These are the so-called dynamic phases of flight, when the vehicle and crew are gaining or losing huge amounts of energy over a short period of time. It was a failure of an O-ring seal at launch that had killed the crew of *Challenger* in 1986 and a damaged heat shield in one of *Columbia*'s wings that had destroyed it and its crew during re-entry in 2003.

Just as the heat assaulting *Soyuz* from the outside was at its fiercest, a red lamp began to flash on the control panel. It was a warning light, telling them that something in their systems had failed and that the vehicle was switching to an emergency backup procedure: ballistic re-entry. It meant that they were plunging inelegantly through the atmosphere – like the cricket ball rather than the frisbee. But Soyeon found this strangely reassuring. The *Soyuz* capsule was designed for this. The ride would be rough but they should still arrive safely.

After what had seemed like an eternity to Soyeon, the violent buffeting stopped and she felt a jerk as the parachutes opened above them. Unsure of what had happened, they checked their systems. It was at this time that they noticed something that looked like smoke coming from beneath one of the panels. In the cramped space of the *Soyuz* capsule, nobody could be sure of what they were seeing but the cloud seemed to hang around Soyeon. With minutes left to go before the descent to Earth was complete, the crew's fears turned to the possibility of fire.

Fire in the confines of the *Soyuz* capsule would be devastating. The crew decided to power down the electrical systems. The re-entry had after all been hotter and harder than expected; perhaps something had overheated and caught light.

Soyeon, however, wasn't convinced. As part of her PhD she had worked daily with liquid nitrogen and liquid oxygen. To her this 'smoke' looked like the vapours from a cryogenic system. Yuri asked her if she was absolutely certain. 'Yes,' she insisted. Reassured, the crew turned their systems back on a short time before landing, but by then they were more than two hundred miles off course.

The capsule hit the ground hard, bouncing before it came to rest on its side in the Kazakh Steppe, far from the intended landing site. The crew unbuckled their straps and crawled out, where they were met by a small group of nomadic tribesmen who initially couldn't understand where Soyeon and her colleagues had appeared from or how they had arrived. Yuri flicked on a satellite phone and called in their position. It was cold – cold enough for their breath to frost the air – and they would have to wait more than an hour before their rescuers reached them. But Soyeon was once again back on Earth and safe.

~

For orbital flight, it is engineering and not clever adaptations or augmentation of physiology that saves lives. The nature of space flight is such that in its most dynamic phases the resilience of our physiology, and its ability to adapt to the physical extremes of the Earth, are utterly irrelevant. The reliance upon artifice is so complete that any significant failure is met with the death of the entire crew. There has never been a mishap in space flight in which only part of a crew has been injured or killed. For every accident the same has been true. Either everything works and everyone lives or it doesn't and everyone dies. And so the first destination on the way to the final frontier is not about our ability to adapt physiologically. It is about the safety of the evolved engineering solutions.

~

Ten days after the launch, I stood in darkness at the shuttle landing facility, swatting mosquitoes and straining my eyes in vain to try and catch sight of *Atlantis*. The last mission of the space shuttle era would land at night, cruelly close to daybreak. We'd be lucky to see anything at all, but we had to come anyway. Somewhere above the crew of STS-135 were on their way home.

A double sonic boom overhead heralded *Atlantis'* arrival. She was circling now, on final approach to Kennedy Space Center, falling unpowered back to Earth, her fuel gone, nearly all of her energy spent. We caught a glimpse of her as she flew through searchlights near the end of the runway before she touched down out of sight. It's not how I had imagined it. I thought that the last space shuttle would land in a blaze of illuminations, and trundle triumphantly across the tarmac, trumpeting the end of an era. Instead, *Atlantis* darted furtively from cover to cover in the half light. Gone almost as soon as she'd appeared, vanishing into what remained of the night like a mythical creature.

It had been half a century since Yuri Gagarin first ventured into orbit aboard *Vostok 1*, in a mission lasting an hour and a half. In those fifty years, the Russians and their American counterparts had learnt to work together in low Earth orbit, transforming it into a staging post for still more ambitious feats of exploration. People now permanently lived and worked in space. Low Earth orbit could be visited not just by trained astronauts like Soyeon, but by paying customers. It was time to set sights on new destinations.

CHAPTER 8
MARS

October 1997: the author floating aboard NASA's KC135 weightless training aircraft, better known as the 'Vomit Comet'

WHEN I FIRST arrived at NASA in 1997 as a student, they were all about Mars. The human space flight division buzzed with excitement; there was a sense that the agency might really be about to embark on a new chapter of exploration – the next small step.

There was a kind of Mars underground at NASA, a cadre of folk who had long held dear the hope of sending a human crew to the red planet. For them, low Earth orbit and the Moon were pedestrian destinations. Mars was where the action was at; exploring it would be the defining feat of their generation: a long overdue return to the sort of barefaced ambition that had first made NASA famous. A badge had appeared on the lapels of the faithful: a cheap tin circle about the size of a quarter with the words 'Mars or Bust!' in bold red lettering.

We've imagined sending people to Mars since well before Gagarin's first space flight. Wernher von Braun, principal architect of the *Saturn V* launcher which delivered Armstrong and Aldrin to the Moon, laid out his dreams in the 1948 publication *Das Marsprojekt*, the first mature study of what it would take to send humans across the huge void of space that lay between Earth and Mars.

It was a design of startling ambition. In it Von Braun envisaged an armada of ten spacecraft ploughing on towards their destination, crewed by no fewer than seventy astronauts. In this plan he foresaw the need to place nearly forty thousand tonnes of payload in low Earth orbit, providing a platform of booster stages with which to launch his Martian flotilla.

Von Braun's plan was of course too fantastic in scale to ever be realised but the kernel of these designs underpinned much of what would follow. The idea that future explorers of Mars would be hurled away from Earth by a brief but violent explosion at the start of their journey, and then left to fall freely through space towards their target, became the accepted template for human missions to Mars.

Throughout the twentieth century Mars continued to drift in and out of our thoughts, appearing almost within reach and yet somehow tantalisingly beyond our grasp. Von Braun's designs envisaged 1965 as the date on which the first humans might arrive at Mars. And since *Das Marsprojekt* more than a thousand different technical studies have been conducted, each making the assumption that Mars lay no more than twenty years in the future. But that is where Mars has remained: always in our future.

Space is not a single destination. Earth orbit, the Moon and Mars are as different in character as the continents of the Earth. So too are the voyages and challenges involved in reaching these locations. Low Earth orbit is about negotiating the violence of launch and the terror of re-entry; about understanding how we should climb out of the well of gravity in which we live, breaking the bonds of attraction created by the mass of the Earth.

Orbital space flight is a furious sprint, with the energies involved barely controlled; an endeavour in which the frailties of human physiology are swamped by the physicality of the propulsive systems. For the pioneers of this age, the ability of the human body to adapt to the extremes of terrestrial environments was largely irrelevant. Dangers were more immediate and dramatic – catastrophic explosions that no one could hope to survive.

Mars presents a challenge of a different scale and character; it's more a marathon than a sprint. The Moon hangs around a quarter of a million miles away from the surface of the Earth. It is a distance we can easily conceptualise: the number of miles the odometer in your car might clock up before the vehicle seized and failed. The Moon, the furthest point from the Earth any human in the history of our species has ever travelled, lies close enough to inspect with little more than the naked eye reachable within four days of space flight.

Mars gets no closer than 35 million miles away. And its position relative to the Earth is always changing, stretching that separation to as much as 400 million miles. To cross that gulf, astronaut crews will have to endure missions drawn out over months and years, spanning hundreds of millions of interplanetary miles, travelling thousands of times further than Armstrong and the Apollo pioneers.

These crews too will have to survive the energies of launch, and those involved in rocketing them away from Earth and towards Mars. But as they fall across the void that separates the two planets they will also have to contend with the silent threat of space and its environment. Here the absence of gravitational load takes on a new dimension, transforming from a novelty into a creeping threat.

~

The term zero-G is a misnomer. Weightlessness in low Earth orbit does not arise because there is no gravity. The gravitational attraction of the Earth doesn't suddenly melt away to nothing just because we venture 250 miles away from its surface. At that altitude, the force of gravity is only modestly diminished: around 90% of its value at sea level. If you were somehow able to build a house on the end of a pole 250 miles long and live in it you might have trouble noticing the change. A dropped glass would still break; climbing stairs would still require effort. There might be something of a spring in your step – you and everything around you would be around 10% lighter – but you wouldn't find your-self floating around from room to room. The weightlessness of orbit is experienced not because of the astronauts' separation from the Earth but because of the way they fall around it.

Weightlessness is something we have all experienced, it's only that our experience of it is generally so brief as to be barely noticed. If you jump up as hard as you can, you might stay in the air for a little over a second. For that time you are weightless.

You could prolong the experience simply by falling further. Imagine standing in a lift on the thirtieth floor of a skyscraper at the moment the supporting cable snaps. From the moment of release until the moment of impact you'd be weightless – a ride of around 300 feet that would last a little over four seconds.

In the same way, astronauts in low Earth orbit find themselves floating because they are inside a spacecraft that is in free fall around the Earth.

But, as I discovered, you don't need a spaceship or a renegade lift to viscerally experience this sort of free fall. Not in France, in any case.

~

Strapped into my seat aboard a modified Airbus, I'm waiting to watch how the French do weightlessness. This is a specialised flight, under the auspices of the European Space Agency's 'DGA Essais en Vol' (literally 'tests in flight'): pilots who specialise in flying aircraft high into the sky in a parabolic arc and then plunging them into a steep dive, pulling out just in time to avoid disaster. At least, that's the theory.

There is a flurry of activity before the start of the parabolas. In place of air stewards we have frequent flyers in tangerine jumpsuits, there to lend a hand if things get rough: the so-called 'orange angels'. People get ready, tweak experiments and position themselves preparing for the next stomach-lurching manoeuvre. 'One minute,' comes the Gallic voice over the intercom, starting the countdown. The scurrying becomes more frantic.

'Twenty seconds . . . ten seconds . . . Pull up!' comes the same disembodied voice. The words are spoken levelly, with no hint of excitement. The person uttering them is at the controls of the plane.

For the next ten to twelve seconds we are pushed into the foam covering the floor of the aircraft. We experience close to twice the normal gravitational load. The burden of my twelve-stone frame is suddenly doubled. I feel like I'm made of lead.

This is nothing compared to the loads that fighter pilots experience during fast turns, but it's more than enough to create discomfort. It's not just toughing out the extra weight; this manoeuvre is perfect for confusing the hell out of the delicate system of accelerometry in your inner ear.

'Thirty,' calls the pilot in the same level tone of voice, narrating the angle of climb now instead of time. We are on our way up to the top of the rollercoaster. That's exactly how it feels, the nervousness, the anticipation, the excitement – and that's not far off what it is. Only this ride is 25,000 feet high and will repeat itself thirty times in the next couple of hours.

'Forty,' comes the voice. 'Inject.' And then, in one of the most effective rapid weight-loss programmes the world has ever known, I go from being twenty-four stone to weighing nothing.

They refer to the point at which the plane begins to fall away from you as rapidly as you are falling towards it as 'injection'. And it does indeed feel like you've been injected into an alternate reality, one in which the normal laws of physics have been briefly suspended. Around you people and things tumble weightlessly, with no respect for the concepts of up or down. The effects of gravity are suspended here. All those dreams you ever had of flying? Well this aircraft makes them come true for twenty-three seconds at a time.

The Airbus drifts over the top of its parabolic arc, its lift balanced perfectly against its weight, thrust throttled to match drag.

'Thirty . . . Twenty . . . Pull out.'

After hanging effortlessly in mid-air for one second, I'm smashed back into the deck. The phrase 'back down to Earth with a thud' could have been invented for the experience of parabolic flight.

Glancing outside I see the wing tips flexed, two metres out of their normal position, like a tensioned bow. More alarming still, a steady trickle of fuel escapes along the wing edges. Swallowing hard, I turn back to the cabin.

The 1.8G load pours on. People's faces appear to age visibly as gravity takes on the skin's elastin and wins. I'm lying on the deck, still managing a smile, when I catch sight of one of the other passengers, head buried in his arm, sweating beads.

One of the orange angels asks him if he's OK. He shakes his head vigorously. He's manhandled to the rear of the plane. There's a fumble for a sick bag and the familiar sound of retching. It's not for nothing that they call this the Vomit Comet.

～

In our daily lives gravity is that pedestrian physical force that keeps us glued to the ground. We don't think of it as something that shapes our lives. Our bodies are set up to allow us to move within its field of attraction without too much effort, so much so that we barely notice it. You have to go out of your way – climb a cliff face or jump out of a plane – before it starts demanding your attention. But we are constantly sensing the effects of gravity and working against them – largely unconsciously.

We are, for example, equipped with anti-gravity muscles – those groups that work against the Earth's force of attraction to keep you standing upright. To get an idea of which groups these are, imagine being on a parade ground with a sergeant major barking at you to stand to attention. Pretty much every muscle you would tense to avoid the prod of his baton is antigravity in function.

Of these the quadriceps, buttocks and calves, along with a group of muscles – the erector spinae – that surround the spinal column and keep it standing tall, are the most important. Without them the pull of gravity would collapse the human body into a foetal ball and leave it curled close to the floor.

These muscle groups are sculpted by the force of gravity. They are in a state of constant exercise, perpetually loaded and unloaded as we go about our daily lives. It is because of this that the quadriceps, the mass of flesh that constitutes the bulk of your thighs and works to extend and straighten the knee, are the fastest wasting group in the body.

Your bones too are shaped by the force of gravity. We tend to think of our skeleton as pretty inert, there to provide rigidity, little more than a scaffold on which to hang the flesh or a system of biological armour. But at the microscopic level it is far more dynamic: constantly altering its structure to contend with the gravitational forces it experiences; weaving itself an architecture which best protects the bone from strain.

And the biological adaptations to gravity don't stop there. When you're standing up your heart, itself a muscle pump, has to work against gravity, pushing blood vertically in the carotid arteries that lead away from your heart towards your brain.

Even your system of balance and co-ordination appears to rely in a fundamental way upon the constant force of gravity, with the otoliths – the organs of the inner ear that sense linear acceleration – using it as a sort of calibrating input.

Life on Earth has evolved over the past three and a half billion years in an unchanging gravitational field. In that context it shouldn't be a surprise that so much of our physiology appears to be defined by, or dependent upon, gravity. Take gravity away, and our bodies become virtual strangers to us.

～

As a medical student you don't take the contents of the middle and inner ear very seriously. The organs within detect acceleration and audible stimuli, gathering information about motion and sound. But they are not considered 'vital', in the sense that they are not required to keep the human body alive. As a result the essential role they play in delivering a finely calibrated sense of motion is often overlooked.

However, like all of the best things in life, you don't really appreciate what you've got until you lose it.

The system of accelerometers in your inner ear, the otoliths and semi-circular canals, are engineered to provide the finest detail about movement in the ever-changing world about you, creating the illusion that you are essentially a stable platform through which the world can be observed as though it were a film made with a steadicam. It is a system that shares its inputs and outputs with the eyes, the heart, the joints and the muscles.

Consider for a moment the act of looking at stuff. Hold a finger up in front of your eyes. Now shake your head left and right as though you were vigorously saying no. The image of your finger remains remarkably stable, doesn't it? Now try keeping your head still and waggling your finger back and forth at the same rate. This time the image is less stable; plenty of blur creeps in.

Keeping an image stable and clear in your visual field is a pretty difficult task to achieve. First you have to focus the image onto the layer of light sensitive cells at the back of your eye called the retina. Now your retina isn't the same all over. At the rear, near the centre, is a cluster of densely packed cells, cone-like in shape, that accounts for less than 1% of the area of the retina. This tiny but all important area is called the fovea, and is responsible for tasks such as reading or studying a picture. This high density of specialised cells resolves the critical detail of a scene and its colours. The rest of the retina, by comparison, is populated by rods – good in low light conditions but rubbish at subtlety. Uninterested in nuance, they're there chiefly to pick up movement in the periphery, to identify a target on which you should focus your attention more closely.

Those receptors report back to a specialised part of the brain called the visual cortex. What's interesting is that although the fovea accounts for less than a hundredth of the surface area of the retina – one voice among a hundred – the visual cortex dedicates 50% of its mass to listening to the super-sensitive discriminating fovea.

All this effort, and we're still talking about a stationary eyeball focusing on a stationary object.

Now let's start shaking things up a bit. Imagine that the thing you're looking at is no longer stationary but is instead moving. As it moves you have to rotate your eyeballs to keep its image focused in the right spot. Once it reaches the point at which you can't track it with your eyes anymore, you start to move your head too.

Now you have two spheres, capable of rotating independently, carrying a lens system that is trying to keep the image of a moving object sharp, on an area at the back of the eye that is only a few millimetres across.

It is the slaving together of the accelerometers in your inner ear, the muscles that rotate your eyeball and those that turn and tilt your head, that allows you to achieve this remarkable feat.

Now imagine that the system doesn't work and that the stable image of the world you take for granted is replaced by a gently oscillating, nausea-inducing scene from which there is no escape. If you've ever suffered from seasickness, imagine the worst possible episode of that, on a ship that you will never be allowed to leave, and for which the rolling seas will never calm down. That's what it feels like when the organs of the inner ear malfunction. And that can be caused by disease, drugs, poisons and – as it turns out – the absence of gravity.

◈

Weightlessness may sound like fun, but the majority of rookie astronauts feel sick in the first forty-eight hours of space flight. Anti-emetic medications – those drugs that act to combat feelings of nausea – are among the most commonly prescribed during NASA space flights. The undesirable effects don't stop there.

Deprived of gravitational load, bones fall prey to a kind of space-flight-induced osteoporosis. The balance between the populations of cells responsible for laying down and removing bone is lost, and so bones become less dense and more prone to fracture. And since 99% of your body's calcium is stored in the skeleton, as it wastes that calcium finds its way into the bloodstream, causing yet more problems.

Hypercalcaemia – a pathological state in which the levels of calcium in the blood are raised – is famous for causing a tetrad of clinical problems. Constipation is the least of these, followed by pains in the long bones. More seriously, renal stones can form, blocking the route from your kidneys to your bladder, causing excruciating pain. And finally there is the possibility of psychotic depression. This list medical students remember as: bones, stones, abdominal groans and psychic moans. All four are problematic when you could be two years and more than four hundred million miles from your closest GP.

And it's not just your bones that waste away. Muscles do too – the anti-gravity groups at an alarming rate. In experiments that charted the changes in the quadriceps of rats flown in space, more than a third of the total muscle bulk was lost within nine days. More interestingly still, astronauts' muscle fibre switches from slow twitch– the efficient fatigue-resistant type suited to marathon running – towards the fast-twitch variety that a sprinter might prefer.

Meanwhile, the heart and its system of vessels, deprived of the need to work against the force of gravity, become deconditioned. The act of space flight enforces a sedentary existence on otherwise well-exercised physiological systems, slowly taking athletes and turning them into couch potatoes. For the cardiovascular system, the set of finely tuned reflexes that on Earth constantly cope with changes in posture sharply deteriorate during extended space flight.

Picture yourself lying on the sofa, watching back-to-back movies. The doorbell rings and you spring to your feet; your cardiovascular system is forced to make a sudden alteration. Having gone from lying to standing, the blood in your body now suddenly tries to pool in your lower limbs, reducing the volume that returns to the heart and, as a consequence, the force with which it beats. In addition, the blood that was lazily flowing between your heart and brain along your carotid arteries is now trying to travel vertically against the pull of gravity.

Combined and unopposed, these changes will leave your brain deprived of an adequate blood supply and you unconscious on the floor.

All that stands between you and that fate is a reflex that senses the drop in pressure in the carotid arteries and tells the brain to increase the rate and force of contraction of the heart, while simultaneously constricting peripheral blood vessels to restore blood pressure. This primitive reflex is all important. Without it, you'd end up lying in a crumpled heap every time you stood up too suddenly.

This is what we see in astronauts returning from long-duration missions aboard the space station. Asked to stand still and upright for ten minutes, a significant fraction are unable to do so without feeling faint. This we call post-flight orthostatic intolerance – an inability to maintain an upright posture.

And the impairments don't stop there. There are other, less well understood alterations. Red blood cell counts fall, inducing a sort of space anaemia. Immunity suffers, wound healing slows and sleep is chronically disturbed.

In short, most astronauts return from long-duration space flight – missions of more than six months' duration – in a temporarily diminished state: sleep deprived, their cardiovascular system de-conditioned, their muscles and bones weakened and their hand–eye co-ordination impaired. As blissful as the experience of floating around might appear, it erodes the body's ability to function when challenged again by the force of gravity.

When astronaut crews arrive back on Earth they are met by a support team which includes nurses and physicians, and they are spirited away to recuperate from the experience. And even then, with all the care that the assembled terrestrial recovery forces can muster, there are still incidents. Returning crew members have been known to vomit at celebratory banquets, collapse in showers or run their vehicles off the road because of transient disorientation.

Others, forgetting that they have returned to a world ruled by gravity, drop expensive equipment or fragile gifts, having got used to the idea that released objects float rather than sink to the floor. Back at home, one astronaut reportedly got out of bed to change his infant son's nappy and stood for a while wondering how he might Velcro the baby to the cot while he searched for some wipes.

The problems of space flight are principally those of re-adaptation to a world in which gravity is the shaping force. Re-acclimatising to that, both physically and psychologically, is a challenge. On return to Earth, astronauts are carefully monitored while their bodies re-adapt. But on a mission to Mars they'd arrive and be entirely on their own.

The crews that arrive at Mars would do so after six to nine months of flight, and experience many if not all of these problems. There they would have to perform the most challenging landing in the history of human space flight. The communication delay between Earth and Mars might be up to twenty minutes. In that moment of touchdown they would be truly alone. Assuming they land safely – and remember that around 50% of everything we've thrown at Mars has crashed or disappeared – they'd then have to leave their vehicle to walk to the pre-prepared habitat. That habitat might be up to half a kilometre away.

And that's assuming they even make it that far.

~

It's worth briefly considering what it takes to get to Mars. The term 'space flight' is something of a misnomer. Human-rated spacecraft don't really fly through space. Their rocket motors fire for only a few brief minutes at the start of the journey, throwing the vehicle and its occupants towards their intended target, like a medieval ballista hurling a missile at the walls of a castle. The spacecraft have their own rocket motors and thrusters, but these are far less powerful than the launcher that set them on their way. Once they're travelling, only subtle course corrections can be made. So astronauts on their way to their destination are engaged in an activity that might more accurately be described as 'space fall'.

While the vehicle and its crew are busy falling across space, Mars is out there somewhere in the darkness, tearing around its elliptical orbit at a little over fifty thousand miles per hour. Mars' journey around the Sun takes 687 days. Earth completes its orbit in 365.25 days, moving at around seventy thousand miles per hour, which leaves the two planets constantly changing their relative positions in the sky.

This has consequences. It means that you can't decide to go to Mars any time you want. You have to wait for precisely the right opportunity, launching from low Earth orbit at exactly the right time, so that Mars is there when you arrive. And the same is true upon your return.

Despite these restrictions there are as many different recipes for getting to Mars as there are for the perfect chicken soup. Mission architects have to juggle propulsion systems, trajectories, vehicle velocities and atmospheric entry strategies and trade these against payload mass and crew size in an attempt to design something realistic in terms of risk and cost. They have to decide, for example, between exotic deep space manoeuvres – that might use the orbital energy of Venus as a slingshot to propel vehicles on their way to and from Mars – and prosaic but potentially safer journeys.

But in the end all of the mission designs boil down to two broad scenarios: those that see you arrive and stay on Mars for a few weeks and those that leave you on the surface of the Red Planet for more than a year. These are the so-called 'Short Stay' and the 'Long Stay' mission architectures for Mars.

For the Short Stay missions crews would travel for close to nine months to get to Mars. But once there they could then take advantage of an early opportunity to return to Earth which would arise between 30 and 90 days after their arrival. This, after having spent close to nine months in flight, would be like flying from London to New York, milling around in the gift shop at JFK for an hour, and then flying straight home. But it has the advantage of shortening the total mission duration to far less than 24 months.

For the Long Stay missions you can get to Mars a little faster, closer to six months than nine, but in this case the elliptical movements of the planets mean that you don't get a chance to come home again for something like eighteen months.

That means you'd spend at least a year travelling and a year and a half or more on Mars. That mission would approach three years in duration – all of which would be spent weightless or working in the reduced gravity of Mars.

There *is* a number of formidable problems that accompany

missions of such duration. The first is life support. How do you invent a system that can keep a crew of four alive for nearly three years? For the International Space Station, breathable oxygen is generated by electrolysing water: using a current to decompose it into hydrogen and molecular oxygen. This requires a steady supply of water, which is conveniently resupplied from Earth via the Russian *Progress* vehicle: a kind of automatically piloted, space-age delivery truck. The carbon dioxide that would otherwise accumulate is scrubbed out using molecular sieves. Spare components for these devices also need to be resupplied aboard the *Progress* vehicle, along with food for the crew.

But there is no easy way to resupply a team travelling to Mars and so a number of ingenious solutions to this problem have been proposed. One involves a grow-your-own approach to life support and nutrition.

One of the experiments under way when I first visited Johnson Space Center as a student in 1997 was exactly this. Plants respire photosynthetically, by taking in carbon dioxide and generating oxygen and water. It turns out that if you grow ten thousand wheat plants you can generate more than enough oxygen for one person to breathe while removing the human waste gases of carbon dioxide. Better still, you have a partial source of nutrition. For a while the Space Center had a team of four volunteers locked up in a hermetically sealed tube, subsisting pretty independently on this self-regenerating, hydroponically grown life-support system. And that's all great – until you factor in the possibility of crop failure.

Another solution, discussed at a European Space Agency human space exploration symposium, would be to grow vats of algae, which might be easier to sustain than wheat, and would also provide a source of protein. Between that and the wheat plants you could get halfway to a diet of pizza-like food – bread coated with flavoured algae – and massively reduce the weight and volume of the food and life-support apparatus required for a Mars mission.

After that conference I remember listening wide-eyed in the bar while an excitable Frenchman who specialised in the field of regenerative life support told me how it might work; going so far as to explain the recycling of urine and the use of faeces as a source of fertilisation.

'You see,' he shouted above the din of the bar, 'these people who go to Mars, they will literally 'av to eat their own shit.'

~

If that hasn't put you off the trip already, then consider the radiation hazards. As far as anyone can tell, the background radiation you would be exposed to while travelling between Earth and Mars should be within safe limits – unless there's a solar flare.

These giant eruptions of plasma from the surface of the Sun are accompanied by an intense shower of high-energy particles that rain through space. For the astronauts and cosmonauts operating in low Earth orbit, within the cage of protection provided by the Earth's magnetic field, this presents little problem. The charged particles are caught and trapped by the lines of Earth's magnetic flux, depositing their energy more or less harmlessly, well away from the human crews.

But for a vehicle venturing outside the Earth's immediate neighbourhood there is no such protection. A solar flare is like a neutron bomb going off next to you. Energetic particles – charged helium nuclei, neutrons, protons and their like – would pass through your body, wreaking havoc and irreversibly damaging cells. Such an exposure would be like taking the DNA blueprints of each cell, shooting cannon balls through them and then trying to build something based on the information that remains. The resulting structures would be dangerously unstable and prone to malfunction.

The fastest proliferating cell populations would be worst affected: hair follicles, skin and the lining of the gut. The rapidly dividing cells of the bone marrow too would fall victim. With blood cells decimated, the sufferer would be left anaemic – short of platelets to help clot blood and bolster the immune system. This explains the familiar depiction of acute radiation sickness: hair falling out in clumps, diarrhoea, bruised skin and bleeding gums. Without a shield, it would be impossible to survive such an exposure.

To make matters worse, solar flares arise sporadically, and we're about as good at predicting them as we are at forecasting the British weather. And there's no straightforward way of combating their effects. Building a ship coated with lead wouldn't help – even if you could find a way to lift that mass into orbit. Lead and other heavy metals are great at shielding against X-ray radiation and lighter particles, but when it comes to highly energetic heavy particles they are worse than useless. Massive particles, arriving at close to the speed of light, would smash into the atoms of a metal shield and scatter them like a cue ball hitting a billiard pack. These scattered atoms would then give rise to secondary radiation, as deadly as the particles they were supposed to shield against.

One possibility lies in building a sort of bomb shelter in the spacecraft; an area more resistant to the radiation storms brought by a solar flare. This you could shield, not with layers of metal, but with a jacket of water. It turns out that water is very good at attenuating solar particle radiation. But this is pretty speculative. When it comes to the radiation hazards of a human mission to Mars, if you ask the experts, they tell you that we simply don't yet know enough.

∾

Even if we figure out a way to negotiate the radiation and build a life-support system that is at least partly regenerative, we keep getting back to the most elemental problem: having to contend with the absence of gravity. The longest mission in human space flight history was 437 days 17 hours 58 minutes and 16 seconds; and was completed by cosmonaut Valeriy Polyakov aboard the

Russian space station *Mir* between 1994 and 1995. By all accounts he arrived back on Earth in reasonably good health, but it is far from clear that this would be true of all space explorers.

Polyakov is in an exclusive club. Around five hundred people have flown into space. Of these only ten have flown for more than 200 days and only two for more than a year.

Most of our experience in the field of astronautics involves missions of less than two weeks' duration. The impairments seen in crew members who have flown for between three and six months are significant and tend to vary from individual to individual.

A range of countermeasures to combat the effects of longer missions is available to astronaut crews. These include medications, special diets and regimes of resistive exercise. But while they have gone some way to mitigating the consequences of human space flight, none appears uniformly effective.

It is because of this that the idea of generating artificial gravity has surfaced time and time again. The concept is not new. The earliest rocket scientists realised that their crews would experience weightlessness and that this might be problematic, even if they could not predict all of its effects.

In 1923 Herman Oberth proposed a solution: a vehicle tethered to a counterweight that would spin end over end like a twirling baton, subjecting the occupants to an artificial gravitational load as it went. It's the same load we feel on spinning fairground rides, the force that pins us against the side of our waltzer car.

So far, so good. But the problem with artificial gravity lies not in the underlying physics of the idea but with engineering a rotating vehicle capable of the feat. Here design is narrowly constrained by the biological frailties of the astronaut crew.

The force of artificial gravity generated by a rotating vehicle depends upon the radius of the vehicle and its rotation rate. To generate enough force it must either be small and spin extremely quickly, or large and allowed to spin more slowly.

Everybody differs in their tolerance to fairground rides: some people can be spun at head-snapping rates without apparent ill effect while others feel sick just watching the thing go round. This, again, is down to the apparatus of the inner ear: detecting rotational accelerations, trying to make sense of what is happening and expressing displeasure through the vomiting centre if it cannot. But if the rate of rotation is kept slow enough, to four revolutions a minute or less, everybody can adapt to the motion in time.

With that requirement fixed, the radius of rotation necessary to produce a force of 1G – equivalent to the load you would feel at the surface of the Earth – can be calculated. It leaves you with a vehicle around 125 metres across – coincidentally about the same size as the London Eye. And if the thought of something of that size whacking around four times every minute seems daunting, imagine building a vehicle of that scale and then launching it into space.

NASA did more than imagine. In the nineties, Kent Joosten and a team of engineers at Johnson Space Center came up with a broad-brush design for an artificial gravity vehicle that might actually work. This returned to Herman Oberth's original idea of a tether between a crew habitat and a counterweight. In Joosten's design, the module and its counterweight were separated by an ingenious, ultra-light, liquid-crystal pylon structure. This could be compressed and stored during launch from Earth and then deployed after the vehicle had arrived in orbit. The whole thing would then tumble end over end all the way to Mars, with the crew living in a module about the size of a four-bedroomed house under conditions that approximated terrestrial gravity.

Joosten's artificial gravity study represents the most mature technical approach to the subject so far seen. There are, however, a number of significant problems to be overcome before such a vehicle design can be realised. It presents an entirely new paradigm in our concept of what human space flight is and this has in part contributed to a reluctance to embrace or further investigate the idea.

Among the hundreds of studies that have considered how best to get to Mars, nearly all of them have involved smaller, simpler vehicles of the type which took us to the Moon. But there is a way to deliver artificial gravity inside such spacecraft, even if the vehicle itself can't be spun.

In our daily lives our bodies do not experience constant gravitational load. When we stomp up and down stairs our joints become shock-loaded, with regions of our skeleton transiently experiencing up to three or four times the gravity they would at rest. When we lie down to sleep, the long axis of our body is more or less perpendicular to the force of gravity, and our skeleton, cardiovascular system and anti-gravity muscles are left unloaded. This quasi-weightless state quite closely resembles the weightlessness of space flight. Indeed, when researchers want to mimic the effects of microgravity here on Earth they simply send a bunch of people to bed.

So on Earth our physiology is maintained by only intermittent exposure to gravitational load – the standing up and stomping around we do during the day. And even that isn't constant. From this realisation grew the idea that we might prescribe gravity like a drug, giving it in short but large doses. Cue the short-arm centrifuge as a countermeasure to the effects of weightlessness. Instead of building a spacecraft as big as the London Eye and rotating it slowly, you could build a much smaller spinning device, rotate it very quickly and pack that inside a conventional spacecraft module.

If you do the maths on this, a centrifuge with a radius of three metres would have to spin around forty times a minute to generate a load of around 3G at its edges. This bizarre regime of loading might nevertheless be enough to protect the body from weightlessness. Better still, it can be administered in short doses: as little as an hour a day might be sufficient. And with this knowledge in hand NASA went out and built one.

~

Somewhere in a NASA laboratory in Galveston, the ceiling spins around above my head, revolving forty times a minute. I keep my head straight, eyes fixed on the screen mounted above, about three feet from my face.

Deep within my inner ear are tiny cells with hair-like protrusions that waft in a gel, like blades of grass standing vertically, set in a plate of jelly. These are part of my vestibular system and exist to detect acceleration in the world around me. The more the jelly leans over, the more the blades of grass bend, and this triggers the firing of the hair cells. Right now they're struggling to make sense of what I'm being put through.

The set of hair cells in my semi-circular canals, the organs that detect rotation, are screaming, firing constantly with the whirling of my body. My brain got bored of listening to that quite some time ago and has decided to ignore their messages, leaving me feeling almost comfortable. But it's a precarious state. There is profound conflict between what I'm seeing and what I'm feeling. My vomiting centre – which is wired in to the same box of tricks that senses acceleration – is at this instant just about managing to stay quiet. I have to keep my head dead centre to maintain that status quo. If I start jerking it around I'll be vomiting in seconds.

I'm wearing a headset with a microphone on. A researcher in the control room watching the camera feed asks me if I'm still OK. I tell him that I am. Another voice from the control room bombards me with a few more questions and then asks me if I wouldn't mind turning my head to take a look at a piece of kit he's worried about on my right-hand side. I tell him that I'm not falling for that one. Somewhere off mic there's an evil chuckle.

I've been here now for half an hour; there are still another thirty minutes left. I am lying on my back on this experimental device: a centrifuge small enough to be accommodated in the module of a spacecraft on its way to Mars.

It looks, at first glance, like an instrument of torture. A pair of arms, each one just about long and wide enough to accommodate an adult lying on their back, sprout from a central column. There are harnesses and straps to stop you flailing around, probes and monitors designed to extract information from you. The whole thing can rotate at a stomach-wrenching rate. If Tomás de Torquemada had invented a fairground ride it would look something like this.

The apparatus is there to interrogate human physiology, to determine how it will respond to this insult. ECG electrodes are glued to my chest, an automated blood-pressure cuff inflates periodically and a probe monitors the oxygen in my bloodstream.

This is a device for generating artificial gravity – or at least an artificial gravitational load. The forces generated when the machine rotates force me out, trying to fling me towards the walls of the room. I'm stopped from doing so by a plate at my feet. As the centrifuge spins up I get heavier against that plate. At full tilt the force on my body below my waist is between two and three times that of normal gravity. Higher up my body, where the speed of travel is slower, the load is less. It means there's a gradient of force across my body that builds steadily from head to toe. This gives the illusion that I'm lying with my back arched, making me feel like I'm engaged in some sort of limbo dance manoeuvre.

As I settle down into it I begin to feel more comfortable. Comfortable enough to begin to get bored. If I don't move my head around the whole experience is quite doable, almost relaxing. A voice crackles into my headset.

'How're you doing?' asks the researcher. I tell him I'm fine. 'We can stick something on the screen if you're getting bored.' He fumbles around in the control room and slides a DVD into the player. A Harry Potter film springs into view on the screen above me and, all of a sudden, whirling in the darkness, in front of a small glowing screen, it feels no more abnormal than watching an in-flight movie on a long-haul flight. And I begin to think that this could be an OK way to get to Mars after all.

~

Artificial gravity is one of those things that people tend to dismiss with a snort if they don't know much about it. It remains unclear how long humans can be deployed in space without suffering serious medical consequences, but it is unlikely that we can endure weightlessness indefinitely and maintain acceptable health. If we are to continue to push out into space then, at some point, artificial gravity machines – compact torture chambers or giant twirling batons – will have to play a role. This is a natural progression. We take everything else with us into space:

our light, our heat, our food and water; we even take our atmosphere. At some point it seems certain that we'll take gravity with us too.

There's a job of work to be done before that can happen. We are unsure of the prescription, of how hard and how fast you would need to spin a crew member to protect them from the ravages of weightlessness. Neither do we know how much protection the partial gravity of Mars will provide, if any.

And it is unclear what such a system might do to the inner ear. Early results from NASA's Artificial Gravity Pilot Project suggested that the heart and muscle might be usefully protected in this way. And it would be surprising if bone didn't benefit too. But the inner ear and its organs of accelerometry are a different story. This strange rotational input might, over time, lead to maladaptive changes that might worsen their function. On the other hand it might prove highly protective. Sadly it doesn't look like we'll find out the answers any time soon.

In 2009, just as the artificial gravity project was ready to enter a more comprehensive phase of investigation, a series of budget cuts tore through NASA. The strategy that would have seen the short-arm centrifuge investigated thoroughly on the ground and then made ready for flight aboard the space station was canned. It isn't the last we've seen of this; as one of the investigators quipped: 'Artificial gravity is an idea that comes around and around . . .'

About the same time a new vision was set: one which prioritised a return to the Moon over a first human mission to Mars. And once again the Red Planet receded into the future.

CHAPTER 9

FINAL FRONTIERS

Simeis 147: a supernova remnant some three thousand light years from Earth. These are the remains of a star that has reached the end of its life, after it has been destroyed by a massive thermonuclear explosion at its core. The energy of that catastrophic event is enough to allow heavier, more exotic elements – of the type upon which life depends – to be created. This, in many respects, is where life in the Universe begins

IN 1917, PRIVATE James Hudson cut an unimpressive figure on the Western Front. He was small but wiry in build and, by his own count, 5 feet 4 inches tall. Overloaded with his full complement of kit, he struggled to clamber in and out of the trenches, even when the German guns weren't trained on him and his pals.

On his first day on the front lines of the Great War, he had a nearly lethal mishap with a hand grenade. Standing in the safety of his own trenches he and a small team of men were practising hurling the devices far enough away to avoid shrapnel injuries.

The seventeen-year-old pulled out the pin and hurled the small metal pineapple as hard as he could. But it strayed off course, colliding with the top of the parapet wall and rebounded back. It fell at his feet with the fuse inside burning down, counting off the seconds before it reached the explosive. There was a moment of panic before he and the rest of the bombing party scattered into the zigzag maze of the trenches, safely out of the way of the blast.

Stumbling across the shell holes of the Western Front, and fumbling the occasional explosive device, the young James Hudson was nevertheless at the peak of his biological fitness.

Physiologically he would never be better than he was in that war. Damage suffered by his body, through disease, accident or the wear and tear of everyday living, was addressed promptly and definitively.

The stem cells of his body were capable of stunning feats of regeneration; his immune system was robust; his body boasted huge physiological reserves. He could run faster, fight harder and survive longer in the face of adversity than at just about any other time in his life. But he was already in his second decade and about to enter his third. And the process of ageing would soon begin to gain traction.

The changes were at first imperceptible to the young private; they would have been measurable only in the laboratory by the most discriminating tests. Later that would change.

But Private James Hudson was a born survivor in every sense. He went over the top in the terrible battles of Mons, Arras and later Ypres, staying alive against incredible odds. He also escaped the sweeping pandemic of Spanish Flu that followed the Great War – a disease that claimed the lives of up to a hundred million worldwide.

Throughout his life he continued to defy every expectation, living in three different centuries, watching a world transform beyond all recognition. And in the final years of his life, after more than a century of adventures and near misses, he finally found himself admitted to Mount Vernon Hospital, under the care of a medical team whose ranks I had just joined as a junior doctor.

~

When Mount Vernon Hospital was first built in the middle of the nineteenth century, it was tuberculosis that stood as the great unmet challenge. It was a disease well described but poorly understood. Physicians could do little more than observe the consumptive horror of the infection as it took hold in lungs and spread to hearts, bones, muscles and brains.

Mount Vernon specialised in its treatment. Built on the top of a hill, boasting ward rooms with open balconies, it represented the cutting edge in Victorian tuberculosis therapy: essentially little more than a plan to expose patients to large volumes of fresh air.

Over its life, the hospital was repurposed more than once to meet the changing healthcare needs of the population, as science and technology continued to redefine the fight against death and disease. It received the casualties of both World Wars, becoming a fully-fledged general hospital with an Accident and Emergency unit during the Second World War. Eventually, with the rationalisation of healthcare in London and its surrounding, it lost its A&E department and became a 'cold site' for the rehabilitation of elderly patients and the treatment of cancer. By the time I arrived, the hospital was over a century old and it seemed fitting that in its twilight years part of its *raison d'être* had become the care of the elderly.

Arriving fresh from nearly three years of acute medicine, with nights spent answering crash calls and pounding down corridors, it looked to me, at first, like the medical equivalent of limbo. A maze of small roads ran from the nineteenth-century buildings at the core to more modern units at the periphery.

These aside, the site didn't look as if it had changed much in the last hundred years.

The elderly care rehabilitation unit was housed in a two-storey prefabricated building, one that had been built at some time as a temporary measure but which had since acquired a more permanent role. It was a place that received patients transferred from bigger general hospitals with more urgent pressures on their beds. The job of this essential – but essentially forgotten – corner of the NHS was to restore its patients to something of their former glory in the hope of getting them home once more.

The nights were quiet and the job offered a break from the cut and thrust of intensive care and A&E, a chance, I thought, to focus on making preparations for my dreaded postgraduate exams.

In the evenings when I was on call, I would tour round the wards just before midnight, scribbling the odd prescription, checking on one or two patients that we were worried about before turning in. There was a side-room on a disused ward with a hospital bed and plastic-covered pillows where you could put your head down with the reasonable expectation that you'd get some sleep.

The practice of elderly care at first felt very alien. As a former physics student, I was always looking for a way to reduce the problems I faced on the wards to something simpler, for systems that would collapse neatly into a few lines of equation and a physiological principle. But here medicine was far less algorithmic. There were, you rapidly came to realise, no quick fixes or easy answers to the medical problems that accompanied advanced age.

Part of what I had liked about astrophysics was the abstraction and the simplicity of the systems under study, systems so invariant in property that you could ask questions of bewildering complexity and have a reasonable expectation of getting decent answers.

In medicine it was the other way around. The human body appeared so unfathomable that we could only ever hope to answer the very simplest questions about the people that we treated. We did stuff largely because it worked. And while statistical methods often told us that our therapies were doing some good, we weren't always able to explain why.

The field of anaesthesia, with its emphasis on the integrated physiology of the human body and its attempts to explain acute changes at the level of first principle, was about as close to the reductionist approach of physics as I was ever going to get. And in the practice of trauma too, one was usually dealing with insults to physiologies otherwise uncomplicated by disease: a single, albeit massive, perturbation in an otherwise stable system. To some extent all of this had felt vaguely familiar.

But when it came to the care of the elderly, the challenge became extreme. Here the underlying physiology of aged patients appeared to have been eroded, leaving them with less in the way of reserve, forever teetering on the brink of instability. Superimposed upon this were layers of chronic illnesses and side effects caused by dozens of drugs – many of which had undesirable interactions.

And on top of all of this were considerations about the proper shape of an individual's life: the state of their home; the strength of their circle of family and friends. For the elderly, the true benefit of every intervention had to be understood and weighed carefully against the considerable risks they presented. The physiology of these individuals was fragile and unforgiving.

Having only ever known acute medicine, with all its urgency, the rehabilitation of the elderly was like learning the rules again from scratch. The biology of the younger patients I had got used to looking after was less nuanced and certainly far more robust.

As heroic as the trauma calls had appeared, they were in comparison like a goal-line scramble in a game of football: urgent and played out in seconds, but always with the possibility that you might recover from your mistakes right up until the last instant.

Being a doctor specialising in the care of the elderly was much more like playing chess. A single poorly considered decision could prove catastrophic. Things happened slowly and in small moves. Sometimes the advance of a pawn was all that was needed. Sometimes retreat was acceptable – even necessary.

There are many stereotypes concerning old age, but I quickly learnt that people grow more, rather than less, different from one another as time passes.

The rehab unit was built on two levels with perhaps thirty in-patient beds upstairs, divided between a ward to the south for the women and one to the north for the men. In the third bed from the door on the men's side was Mr Hudson, who had by this point reached the remarkable age of 103. Now frail and afflicted with pneumonia, he was nevertheless sharp in mind and spirit. But the fact of his survival was perhaps less surprising to him than it was to us, his carers. For if there was one trick that James Hudson had learnt in over a century of living it was how to beat the odds.

~

While the twentieth century brought life-saving innovations, it also gave rise to an array of increasingly violent ways to destroy ourselves and each other.

On 12 February 1898, a gentleman by the name of Henry Lindfield became the first recorded fatality from an automobile accident when he lost control of his two-seater and smashed into a tree outside of Purley. He had been driving downhill at the heady rate of 17 miles per hour.

Almost exactly a month later, James Hudson was born in a mews house in London close to Paddington station. Although the house was within spitting distance of St Mary's Hospital, he was delivered at home, to a coachman and his wife, at a time when more than one in every ten newborn infants died at or shortly after birth.

He arrived in the world at the end of the nineteenth century, before motorways or anything that resembled modern medicine, before the Wright Brothers or Einstein's great theories, at a time when Everest stood unclimbed, hearts were considered inoperable and the maps of the world still boasted a vast uncharted continent of snow and ice to the south.

In the year of his birth, London was a city of cobbled streets and horse-drawn carriages. There was no ambulance service, no welfare state or National Health Service. Healthcare was something that only those of means could afford. Everybody else depended upon simple charity.

In a time before vaccination and antibiotic therapy, infectious disease was the leading cause of death. As the twentieth century approached, a child could, on average, expect little more than forty-five years of life. Around two out of every ten children born in that time were dead before the age of five. Nearly a third did not survive beyond twenty-five years. But that was not to be James's fate.

Young Master Hudson left school at the age of fourteen. A bright and determined boy, he took up an apprenticeship with a dental surgery in Tonbridge, hoping that he might one day gain entry to a medical school.

The Great War intervened – and Hudson bore witness to the terrifying efficiency with which mechanised society could destroy lives. But none of this broke either him or his stride. A year after the First World War ended, he enrolled at Guy's Hospital as a student of dentistry, and by 1928 he had his own dental practice.

Working within a hospital as a dental surgeon, Hudson began to notice an increase in the number of facial injuries and fractured jaws as a result of motor accidents. He noted too that with his dental training and knowledge of the relevant anatomy he was better equipped to deal with such patients than most general surgeons were.

He and several of his colleagues campaigned for the establishment of a new speciality – one which embraced dentistry and surgery in a single field, one specifically for injuries and surgical diseases of the head and neck. This became the field of maxillofacial surgery. The boy who had started life above a stable block at the end of the nineteenth century became a consultant surgeon and one of the founders of a new surgical speciality.

~

A lifetime later – and nearly sixty since the birth of the National Health Service – I work my way around the ward from patient to patient, pushing the trolley of notes as I go. From the end of the bed, James Hudson appears frail. Tucked up in a chair with a blanket on his lap, he is thin and bespectacled, with white hair. The sagging of his features represents the disappearance of elastin, a protein that gives skin its youthful appearance. The lines of skin cells that have marched forward over so many decades continue to do so, only these days they are slightly less well made.

The fibres of his muscles too have changed, shrinking back, losing much of their youthful bulk. Their 'cut', the lines of defi-nition that demarcate each muscle group clearly, has faded thanks to a decrease in his levels of testosterone. And so too has the testosterone. The same apparent weariness of his body's production line that is responsible for the changes in his skin has affected every system.

His spoken words are clear but noticeably less forceful than those of his younger visitors. The muscles that shape his voice, like those of his skeleton, have become weaker with time. The vocal cords cannot be held so consistently in position. They waver now in the same way that the muscles of his arm might were he to lift a heavy weight. And the lungful of air that he expels in order to make those sounds is also smaller. Now that the recoil of his lungs is less powerful – like his skin, their elasticity too has been eroded – breathing out is more of an effort. The capacity of his lungs themselves has shrunk.

The evidence of Mr Hudson's great age goes beyond that which I can see and hear from the end of the bed. His biochemistry is deranged. His kidneys are less impressive: less capable of filtering the volumes of blood that course through them; more vulnerable to insult. The toxins and drugs that they are supposed to remove are cleared more slowly these days.

His heart beats with less force, emptying smaller volumes with every beat. The electrocardiogram, which traces the spread of electricity through that vital organ, shows the occasional missed beat.

When he rises from his chair, he cannot pull himself up to his full height. The weakened bones of his spinal column have over the years given way to the forces generated by imperfect posture. The spine itself curls gently forwards now, causing a permanent stoop.

When – with assistance – he gets to his feet, you cannot help but worry. The muscles of postural control, those which keep him upright, are less capable now. They labour under the control of the brain, which is unconsciously and perpetually making corrections to keep him on his feet. But that foreman too is less competent than he once was.

The intricate system of accelerometry in his inner ear – that which in youth and health can tell an Olympic figure skater when to emerge from a pirouette – is now prone to playing cruel jokes, occasionally providing illusions of motion when it is absent, leaving him unsteady. And the bones on which his flesh and muscle are hung are less dense, more prone to fracture when exposed to sudden force.

This is the physiology of great age, and the frailty that accompanies it is undeniable. But for all of the above, at 103 Mr Hudson continues much as he must have done for most of his life. He is the oldest member of his golf course and a man who, until a month ago, still drove a car.

～

It is entropy that we are up against here. Entropy is that property, common to all systems alive or dead, that sees them tend from a state of order to one of chaos. And once the processes of renewal and replenishment that maintain us in our youth begin to run down we are left open to its ravages.

You can think of the biomolecules that comprise the cells and tissues of your body as though they were thousands of trillions of spinning tops arranged on a table top. In their initial state they all stand neatly ordered – upright and resilient to perturbation. But gradually, as they begin to spin down, they slow and become more unsteady.

Those tops can be re-spun, and prevented from falling over, with a whip. Or those teetering on the brink of catastrophe can be removed completely and replaced with a freshly spun top. Whipping and replacing the spinning tops are processes analogous to the repair and regeneration of biomolecules. It is this perpetual input of energy to the biological system that temporarily staves off the consequences of entropy.

In youth, the process of entropy is held in check by the body's intrinsic system of repair and regeneration. A system – if you like – that is constantly replacing and whipping trillions of molecules into shape. But just as a child eventually gets bored of the toy, so the body eventually begins to abandon the processes of repair and regeneration.

Unabated, the entropy of the biomolecules of which the body is comprised manifests itself as the process that we experience as ageing. According to this model, the state of youth is akin to a forest of spinning tops standing stable, fast and full of energy. Old age sees them slowing down, growing unsteady, oscillating wildly – ready for a passing breeze to knock swathes of them over.

Entropy does not cause disease, nor is it disease, but it leaves an organism vulnerable. And disease takes hold where the system is weakest. For humans that is the cardiovascular system and those cell populations most prone to cancer.

Having made great strides against communicable disease in the opening decades of the twentieth century, we are now up against the limits set by entropy.

In a sense we have transcended the fate of other animals, leaving behind the eternal war between species and microorganisms that operate to kill such large fractions of a population in early life, and – barring accident – are left to behave more like the objects in the physical world around us. Winding down as entropy takes hold. Left to fail the way a star might.

～

Mr Hudson's chest is full of crackles. His cough is worse today; his breathing more laboured. The pneumonia that first brought him to hospital has returned. I am unsure what the kindest intervention might now be.

At the start of the twentieth century, the continent of old age was a destination that stood rarely visited and largely unknown. For nearly the entirety of human history, average life expectancy languished at around thirty years of age. This grim statistic remained fairly constant throughout recorded history, about as true of the ancient Greeks as it was of the Victorians of the nineteenth century.

In the time that James Hudson has lived, life expectancy in the UK has nearly doubled. The oldest person who has ever lived was nineteen years older than he is now when she died at the remarkable age of 122 years and 164 days in 1997.

Jeanne Calment, born in the 1870s, lived for her entire life in the French city of Arles and, as a young woman, once met Van Gogh. It is hard to make sense of her longevity. She smoked, ate chocolate, drank port and, according to reports, wasn't particularly fussed about exercise – hardly the behaviour of an individual attempting to stave off the inevitable running down of her physiological processes.

There are but a handful of people alive today who have approached this great age. Fewer than one in a thousand centenarians reaches the age of 110. Could Mr Hudson pass this extreme frontier? Statistically, the chances appear slim. But we are in *terra incognita* here, living at a time where we are seeing remarkable feats of survival and longevity occur more and more commonly.

We are increasingly aware that a patient's chronological age is not the same thing as their physiological age, and that it's a mistake to underestimate those who've passed their allotted threescore years and ten. After all, John Glenn flew into space aboard Space Shuttle *Discovery* aged 77. Jeanne Calment herself took up fencing for the first time aged 85. And at 102 years of age, Mr James Hudson described himself in the national census as merely 'semi-retired'. As his doctor, I have no direct way of knowing how well preserved Mr Hudson is, only that he is resilient enough to have made it this far.

My patient is arguably less likely to die on this day than he was on any day at Arras, Mons or Ypres. And only slightly more likely to die today than he was in his first year of life. But Mr Hudson is an individual – more different from the population upon which that statistic is based than he has been at any time in his life. He has seen unthinkable revolutions in healthcare, science and technology. He has seen the seemingly impossible achieved over and over again. And for him the only constant through all of that time has been the fact of his survival.

Tucked up in bed fighting pneumonia, he is still in many ways the same plucky private from the Western Front – keeping his head down, knowing only that this is war and that all wars are hard.

~

'Nobody,' as Senator John Glenn once put it, 'has yet found a cure for the common birthday.' But for those fortunate enough to live in the developed countries of the world, the continent of old age is, after two million years of human evolution, suddenly open to all. Equipped with only average luck, assisted by the advances that modern living has brought, the vast majority of us will reach it.

We may find difficulty in perceiving old age as a thing of exploration, but that it is – and one in which all of us today can participate. Neither do we regard it in the same way as we do other unexplored destinations: with expectation, hope and curiosity.

But life is an exploration, and James Hudson is among the greatest explorers of any age, living across three centuries, witness to some of the most significant events of the twentieth century. To him it was all an adventure, and one to be enjoyed until the last possible moment.

∽

Entropy, disease and the complexion of our genes eventually catch up with us all, even those who have walked upon the Moon. On 25 August 2012, Neil Armstrong died in hospital, in the city of Cincinnati in Ohio, having failed to fully recover after cardiac surgery. He was 82. Earlier that month he had stepped onto a treadmill and walked, while doctors monitored the dance of electricity in his heart through electrocardiogram leads.

They perhaps noticed small upswings and depressions in the waveforms scribbled out before them, symptomatic of constrictions in the arteries supplying the muscle of his heart. They would have gone on to map those tributaries in greater detail, delineating the number and severity of the narrowings, before deciding upon a plan. That plan was for Armstrong to undergo a cardiothoracic operation, in an effort to bypass the blockages in his coronary arteries, and restore the supply routes to his heart.

Bypass operations were pioneered in the same decade in which Project Apollo reached the Moon. This type of surgery remains among the most invasive that medicine offers – carving into the chest, isolating the heart from its surroundings and establishing the patient on heart–lung bypass machines – and it comes with attendant risks. This Armstrong's surgical team would have laboured to explain, weighing up the alternatives, making clear what might be won and lost in the endeavour.

Precisely how you stratify risk to a man who commanded the first crew to land on the Moon, or how Armstrong himself perceived it, I do not know. Despite his earlier occupation, Armstrong was no adrenaline junky. Unnecessary risks were, in his opinion, best avoided.

Famously he believed that human space flight ought to involve no more risk than making a milkshake. Though back in the summer of 1969, as the lunar module *Eagle* sank towards the Sea of Tranquillity, running low on fuel, its onboard computers having crashed repeatedly, space flight still had a long way to go to catch up with the safety record of the milkshake industry.

The early days of human space flight and heart surgery were watched by the world in wide-eyed wonder. The risks involved in these pioneering endeavours were so great as to be impossible to sensibly quantify. Deaths were expected.

Today the risk of catastrophic failure during the launch of a human-rated orbital space vehicle stands at perhaps two or three per cent, almost the same risk as that involved in coronary artery bypass surgery. Despite these risks being significant, both have, to some degree, come to be viewed by the public as being within acceptable limits. They have begun to appear almost routine.

The surface of the Moon, like the anatomy of the heart, had been studied for centuries. Both had stood for millennia in full view and yet unexplored. The Moon was reached by an astronaut crew launched across the void, wrapped in a facsimile of Earth's atmosphere. That same approach – of swaddling physiology in systems of artificial life support – was to be the key to successful cardiac surgery.

In the few decades that have elapsed since Scott and Amundsen first marched to the South Pole, we have come a long way. Our expectations of the insults we might survive, in the pursuit of geographical conquest as well as on the operating table, have been transformed. Life has never been safer, never longer lived.

But look closer and the picture gets more complex. Exploration is necessarily a process of trial and error, of taking risks. It appears clear what we have to gain by advancing, so boldly, as clinicians. But we're growing more circumspect about the virtues of physical exploration in science and the physical world, and particularly that which sees us staring out into space at the final frontier. We've begun to wonder if we should continue to boldly go.

∼

The age of human space flight was, without doubt, brought into being by the nuclear arms race of the mid twentieth century. In the 1960s, with the respective nuclear arsenals of the Soviet Union and the United States of America standing ready to bring about their mutually assured destruction, human space flight became a surrogate battlefield for a war that couldn't be fought in any other way.

With the Soviet Union ahead at every point in the space race, there were hard truths for the United States to face. But the reply to Sputnik, Laika and Gagarin was to be Armstrong, Aldrin and Collins. And despite Russia's earlier pre-eminence, the lunar landing in July 1969 somehow gave the United States victory in this bizarre struggle.

Project Apollo and its lunar exploration missions were conceived, built and launched before Kennedy's famous decade was out. This feat appears to grow more miraculous as the years roll by. The Mission Control room that drove those first forays to another world was stocked with slide rules, pocket protectors and Bakelite telephones with rotating dials. To contemporary eyes, it hardly seems possible that the technology of the time could be up to the task of delivering men to the surface of the Moon. In that regard, it is an achievement that stands outside its time, a feat of anachronism.

But perhaps acts of exploration never fully make sense to rational people. They are, after all, about venturing beyond what is known and safe and to be counted on. In this regard, maybe the greatest feats of exploration must always feel anachronistic.

∼

After half a millennium we still remember Ferdinand Magellan, and the straits linking the Pacific to the Atlantic to which he gave his name. We might remember too the extraordinary voyage that saw his flotilla of ships become the first expedition to circumnavigate the globe. The legacy of discovery is what we celebrate. What we recall less clearly is the expedition's legacy of loss.

For Magellan, setting sail from the Spanish port of Sanlúcar de Barrameda in 1519 with a fleet of five ships and 237 crew, the oceans of the world must have seemed as unknown, and presented at least as much threat to life, as the ocean of space that lies between Earth and Mars today.

The expedition endured famine, disease, mutiny and conflict. Magellan himself was slain in the Philippines, in the shallows around Mactan Island, before the circumnavigation was complete. When the expedition finally returned to port in 1522, exactly three years after its departure, only one ship, the *Victoria*, and eighteen of the original 237 crew remained.

Today history recognises this as having been an important feat of exploration, a necessary step towards still greater feats of naval discovery. But to Magellan's crew and the people who lived and worked in the Spanish port into which the *Victoria* limped at the start of the sixteenth century, it could hardly have seemed so.

~

In retelling the story of twentieth-century medicine we often superimpose a narrative of steady progress, when in truth physicians, surgeons and scientists did little more than stumble ahead, as all explorers do, solving and creating problems as they went.

Both Charles Bailey and Dwight Harken endured many failures in the early days of closed heart surgery, resulting in the deaths of a very large percentage of their first cohort of patients. These first efforts at heart surgery were viewed as bizarre and extreme forms of intervention. So too was Archibald McIndoe's practice of subjecting disfigured airmen to dozens of operations, waltzing squares of flesh across their bodies in the hope of reconstructing something of their faces. And Bjorn Ibsen only narrowly convinced his colleagues of the value of artificial ventilation in addressing the devastating effects of poliomyelitis.

Faced either with the scars of the Battle of Britain or the suffocating death of the polio epidemics, both McIndoe and Ibsen could have safely chosen to do nothing. In both cases the threats addressed by their innovation were rapidly neutralised by other means. Aircraft, even combat aircraft, became immeasurably safer through improved engineering. Polio was addressed effectively with programmes of vaccination. Within twenty years of the Copenhagen epidemic that gave rise to the world's first intensive-care units, the spectre of this paralysing illness had all but disappeared from developed countries and today polio is very nearly eradicated from the world at large.

But the intensive-care units that Bjorn Ibsen laboured to create were soon repurposed to treat all sufferers of critical illness – within three decades we could artificially support lungs, hearts, kidneys and even the gut. And intensive care came to underpin the heroic feats of surgery that we have come to expect in the modern age, Dallas Wiens and his transplanted face among them.

Plastic surgery also underwent something of a transformation. The devastation and disfiguration wrought by fires became thankfully rarer over the course of the last hundred years. Its ravages have been replaced by the invasion of cancer. And it is here that the art of plastic surgery, forged in the fire of war, now finds itself most keenly applied. None of these destinations was arrived at through careful planning. But when you strike out into new territory, you rarely know what you're going to discover.

Together, dozens of discrete events in the history of modern medicine came together to ensure Anna Bågenholm's survival. From flying ambulances and cardiac bypass circuits to intensive-care units and reconstructive surgeries: all of it eventually became a continuous chain of survival that took a young doctor from death beneath the ice of a frozen river in Norway to resuscitation and survival in a hospital in Tromso. Anna's survival was one of the unintended consequences of the exploration of earlier epochs. And that, in part, answers the question of why we should explore at all. To be able to explore we must continue to survive. But the reverse is also true: to survive we must explore.

We advance in science, medicine and exploration in fits and starts. There is no real plan – at least not one that anyone has ever stuck to for very long. We happen upon our discoveries largely by accident, making the most of them as and when they arise. We meet disaster in the same way. We explore simply because we must. And that is what makes us human.

SOURCES AND FURTHER READING

ALTHOUGH THIS BOOK covers a century of change in the world of exploration and medicine, it isn't constructed as a historical treatise any more than it is intended to be a textbook of medicine for aspirant physicians.

The narratives within are stories that I was aware of but didn't really know. They are there to provide an insight into the incredible things that we are able to do in medicine and exploration today, and the ways in which we arrived at this point.

But they are a starting point rather than a destination. For those wanting to explore further, I've included the references below, which were used as source material but whose detail it was not always possible to discuss in the depth they deserve.

ICE

I met Anna and Torvind while making a *Horizon* documentary for the BBC. Anna's case is one that has become the stuff of legend among the medical fraternity; one that we have often used to illustrate the fact that in the face of hypothermia it is worth persisting with efforts at resuscitation. But I had never in all of that time referred to the original *Lancet* publication. Torvind and Anna were good enough to come to London and lecture at a seminar I had organised at the Royal Society of Medicine in 2011. There they told their story in its full detail and it is upon that account – and the drenching on the way to dinner afterwards – that the substance of the material in this chapter is based.

My thanks also to the Dezhbod family for permission to include their story and for talking with me about it. Esmail has recovered well from his surgery and has returned to work. His eldest daughter is at college and hoping to go to medical school upon graduation.

It is worth noting that cardiac arrest survival rates have improved significantly in the fifteen years since I first qualified from medical school. I'm hugely grateful to Dr Jerry Nolan, consultant anaesthetist and Chairman of the Resuscitation Council (UK) for answering numerous queries. Changes in treatment protocols and new techniques, including therapeutic hypothermia, have brought significant benefits. And while survival following cardiac arrest remains the exception rather than the rule, junior doctors today thankfully have far less cause for pessimism.

Berdowski, J. Berg, R. A., Tijssen, J. G. P. & Koster, R. W. 'Global incidences of out-of-hospital cardiac arrest and survival rates: Systematic review of 67 prospective studies.' *Resuscitation*. 2010; 81: 1479–87.

Boutilier, R. G., 'Mechanisms of cell survival in hypoxia and hypothermia.' *Journal of Experimental Biology*. 2001 Sep; 204(18): 3171–81.

Gilbert, M., Busund, R., Skagseth, A., Nilsen, P. A. & Solbø, J. P. 'Resuscitation from accidental hypothermia of 13.7 degrees C with circulatory arrest.' *The Lancet*. 2000 Jan; 355(9201): 375–6.

Lane, N. & Martin, W. 'The energetics of genome complexity.' *Nature*. 2010 Oct; 467(7318): 929–34.

Lane, N. *Power, Sex, Suicide: Mitochondria and the Meaning of Life*. Oxford University Press, 2006.

Larson, E. J. *An Empire of Ice: Scott, Shackleton, and the Heroic Age of Antarctic Science*. Yale University Press, 2012.

Mallet, M. L. 'Pathophysiology of accidental hypothermia.' *QJM*. 2002 Dec; 95(12): 775–85.

Haman, F. 'Shivering in the cold: From mechanisms of fuel selection to survival.' *Journal of Applied Physiology*. 2006 May; 100(5): 1702–8.

Nolan, J. *Advanced Life Support*. Resuscitation Council (UK). 5th Edition, 2006.

Scott, R. F., ed. Jones, M. *Journals: Captain Scott's Last Expedition*. Oxford World Classics, 2008.

Solomon, S. *The Coldest March: Scott's Fatal Antarctic Expedition*. Yale University Press, 2001.

Solomon, S. & Stearns, C. R. 'On the role of the weather in the deaths of R. F. Scott and his companions.' *Proceedings of the National Academy of Sciences of the USA*. 1999 Nov; 96(23): 13012–6.

Swinton, W. E. 'Physicians as explorers. Edward Wilson: Scott's final Antarctic companion.' *Canadian Medical Association Journal*. 1977 Oct; 117(8): 61–3, 74.

FIRE

Hundreds of servicemen were injured and then treated by the staff of the Queen Victoria Hospital in East Grinstead during the Second World War; and thousands since. Tom Gleave's story is but one among hundreds of the same era, though as the first and only Chief Guinea Pig, and one of the founder members of the Guinea Pig Club, his seemed well worth retelling. The events of the air battle that led to his injuries have been told several times: in a television documentary made for what was Thames Television (*The Guinea Pig Club* directed by Robert Fleming), in interviews for radio documentaries and in his own book published in 1941 titled: *I Had a Row with a German* after his first words to his wife when she saw his disfigurement and asked what had happened to him. The accounts vary a little from retelling to retelling, possibly because the original source – Gleave's own book originally published anonymously during the war and vetted by the Ministry of Information – modified details to avoid giving away secrets about the Hurricane's vulnerabilities. I have done my best to piece the different strands together. To get a proper sense of just how flimsy and flammable the Hurricane's fuselage was, I visited Hawker Restorations in Suffolk and saw a handful of reconstructed Hurricanes – one of which had been flown by Gleave in combat.

I am grateful for the assistance of Tom Cochrane and Bob Marchant, respectively honorary surgeon and honorary secretary to the Guinea Pig Club for their wisdom and recollections. Mr Cochrane rightly points out that Archibald McIndoe stood on the shoulders of the giants that preceded him, notably those of his cousin Harold Gillies whose pioneering work in the field of plastic surgery is more fully described in the 1920s textbook *Plastic Surgery of the Face* authored by Gillies himself.

My thanks also to Tom Edrich and Bohdan Pomahac for taking the time to talk with me at length about Dallas Wiens' difficult surgery and for reviewing the written material.

Battle, R. 'Plastic surgery in the two World Wars and in the years between.' *Journal of the Royal Society of Medicine.* 1978 Nov; 71(11): 844–8.

Bishop, E. *McIndoe's Army.* Grub Street, 2001.

Fleming, R. *The Guinea-Pig Club.* Thames Television, 2011.

Geomelas, M, Ghods, M, Ring, A. & Ottomann, C. ' "The Maestro": A pioneering plastic surgeon – Sir Archibald McIndoe and his innovating work on patients with burn injury during the Second World War'. *Journal of Burn Care & Research.* 2011 May–Jun; 32(3): 363–8.

Gillies, H.D. *Plastic Surgery of the Face.* Henry Frowde, 1920.

Gleave, T. *I Had a Row with a German.* Macmillan, 1941.

Jackson, D.M. 'Burns: McIndoe's contribution and subsequent advances.' McIndoe lecture, 1978. *Annals of the Royal College of Surgeons of England.* 1979 Sep; 61(5): 335–40.

Matthews, D.N. 'A tribute to the services of Sir Archibald McIndoe to plastic surgery.' *Annals of the Royal College of Surgeons of England.* 1967 Nov; 41(5): 403–12.

Mayhew, E.R. *The Reconstruction of Warriors.* Frontline Books, 2010.

McKinstry, L. *Hurricane.* John Murray, 2010.

Morgan, B. & Wright, M. *Essentials of Plastic and Reconstructive Surgery.* Faber & Faber, 1986.

Mosley. L. *Faces from the Fire*. Weidenfeld & Nicolson, 1962.

Penn. J. 'The reminiscences of a plastic surgeon during the Second World War.' *Annals of Plastic Surgery*. 1978 Jan; 1(1): 105–15.

Pomahac. B., Papay, F., Bueno, E.M., Bernard, S., Diaz-Siso, J.R. & Siemionow, M. 'Donor facial composite allograft recovery operation: Cleveland and Boston experiences.' *Plastic and Reconstructive Surgery*. 2012 Mar; 129(3): 461–7.

Pomahac, B., Lengele, B., Ridgwa,y E.B., Matros, E., Andrews, B.T., Cooper, J.S., Kutz, R. & Pribaz, J.J. 'Vascular considerations in composite midfacial allotransplantation.' *Plastic and Reconstructive Surgery*. 2010 Feb; 125(2): 517–22.

Pomahac, B., Pribaz, J., Eriksson, E., Annino, D., Caterson, S., Sampson, C., Chun, Y., Orgill, D., Nowinski, D. & Tullius, S.G. 'Restoration of facial form and function after severe disfigurement from burn injury by a composite facial allograft.' *American Journal of Transplantation*. 2011 Feb; 11(2): 386–93.

Pomahac, B., Diaz-Siso, J.R. & Bueno, E.M. 'Evolution of indications for facial transplantation.' *Journal of Plastic, Reconstructive & Aesthetic Surgery*. 2011 Nov; 64(11): 1410–16.

Pomahac, B. & Pribaz, J. 'Facial composite tissue allograft.' *Journal of Craniofacial Surgery*. 2012 Jan; 23(1): 265–7.

Pomahac, B. 'Establishing a composite tissue allotransplantation program.' *Journal of Reconstructive Microsurgery*. 2012 Jan; 28(1): 3–6.

Proksch, E., Brandner. J.M. & Jensen, J.M. 'The skin: An indispensable barrier.' *Experimental Dermatology*. 2008 Dec; 17(12): 1063–72.

Scripko, P.D. & Greer, D.M. 'An update on brain death criteria.' *The Neurologist*. 2011; 17(5): 237–40.

Taylor, G.I. & Palmer, J.H. 'The vascular territories (angiosomes) of the body: Experimental study and clinical applications.' *British Journal of Plastic Surgery*. 1987 Mar; 40(2): 113–41.

Wallace, A. F. 'The early development of pedicle flaps.' *Journal of the Royal Society of Medicine*. 1978 Nov; 71(11): 834–8.

'The Hurricane Unveiled', *Flight Magazine*. 12 May 1938; 467.

HEART

The staff of the British Library were particularly patient with me while I researched this chapter. The case of Grey Turner's wounded soldier was detailed in a case report for *The Lancet* in 1940, with a more formal treatise on the subject of gunshot wounds of the heart in 1941 for the *British Medical Journal*. His lecture to surgeons preparing for service during the First World War predates these publications by some twenty-three years.

I was very fortunate to be able to speak with Dr Alden Harken, son of Dwight Harken and himself also a successful cardio-thoracic surgeon. His recollections of the early days of cardiac surgery and the rivalries his father had to contend with were particularly colourful and I am very grateful for the time he was able to spare.

Chaikhouni, A. 'The magnificent century of cardiothoracic surgery.' *Heart Views*. 2010 Mar; 11(1): 31–7.

Cooley, D. 'In Memoriam: Dwight Emary Harken.' *Texas Heart Institute Journal*. 1993; 20(4): 250–251.

Ellis, L. & Harken, D. 'Mitral stenosis, clinico-physiologic correlations, with particular reference to surgical intervention.' *Transactions of the*

American Clinical and Climatological Association. 1948; 60: 59–70.

Fye, W. B. 'Ernest Henry Starling, his law and its growing significance in the practice of medicine.' *Circulation.* 1983 Nov; 68(5): 1145–8.

Gonzalez-Lavin, L. 'Charles, P. Bailey and Dwight, E. Harken—The dawn of the modern era of mitral valve surgery.' *The Annals of Thoracic Surgery.* 1992; 53(5): 916–19.

Grey, Turner, G. 'A clinical lecture on the importance of general principles in military surgery.' *British Medical Journal.* 18 Mar 1916; 1: 401.

Grey, Turner, G. 'Gunshot wounds of the heart.' *British Medical Journal.* 21 June 1941; 1: 938.

Grey, Turner, G. 'Gunshot wounds of the chest. (Correspondence)' *British Medical Journal.* 26 Apr 1919; 1: 530.

Grey, Turner, G. 'A bullet in the heart for 23 years.' *The Lancet.* 1940 Oct; 236(6112): 487–9.

Hadfield, C. F. 'Anaesthetic explosions.' *British Medical Journal.* 9 Aug 1952; 2(4779): 332–4.

Harken, D. E. & Taylor, W. J. 'Diseases of the Cardiovascular System (Surgical).' 2002; 1–34.

Harken, D. E. 'One surgeon looks at human heart transplantation.' 1968. *Chest.* 2009 Nov; 136(5 Suppl): e24.

Harned, C. 'Some practical suggestions concerning the use of alkoform as an anesthetic agent.' *Anesthesia & Analgesia.* 1927 Dec.

Hoyt, D. B, 'Blood and war—Lest we forget.' *Journal of the American College of Surgeons.* 2009; 209(6): 681–6.

Katz, A. M. 'Ernest Henry Starling, his predecessors, and the "Law of the Heart".' *Circulation.* 2002; 106(23): 2986–92.

McCawley, E. L. 'Management of Cardiac Arrhythmias During Anaesthesia.' *Canadian Anaesthetists' Society Journal.* 1955 Apr; 2(2): 137–41.

Naef, A. (2004). 'The mid-century revolution in thoracic and cardiovascular surgery: Part 5.' *Interactive Cardiovascular and Thoracic Surgery*. 2004; 3(3): 415–22.

Proksch, E., Brandner, J. M. & Jensen, J. (2008). The skin: An indispensable barrier. *Experimental Dermatology*. 2008; 17(12): 1063–72.

Selzer, A. & Cohn, K.E. 'Natural history of mitral stenosis: A review.' *Circulation*. 1972; 45(4): 878–90.

Sellors, J.H. 'The genesis of heart surgery.' *British Medical Journal*. 1967; 1(5537): 385–93.

Sinclair, C.M. 'Modern anaesthetic machines.' *Continuing Education in Anaesthesia, Critical Care & Pain*. 2006; 6(2): 75–8.

Søreide, K., Petrone, P. & Asensio, J. A. 'Emergency thoracotomy in trauma: Rationale, risks, and realities.' *Scandinavian Journal of Surgery*. 2007; 96(1): 4–10.

Symbas, P.N., Justicz, E.G. & Justicz, A.G. 'Quantum leap forward in the management of cardiac trauma: The pioneering work of Dwight E. Harken.' *The Annals of Thoracic Surgery*. 1993; 55(3): 789–791.

Waisel, D. 'Norman's war.' *Anesthesiology*. 2003; 98: 995–1003.

Waisel, D. 'The role of the Second World War and the European Theater of Operations in the development of Anesthesiology as a physician specialty in the USA. *Anesthesiology*. 2001; 94: 907–14.

Willan, R.J. 'George Grey Turner.' *Annals of the Royal College of Surgeons of England*. 1951 Oct; 9(4): 274–6.

TRAUMA

The Styner family's story is a prime example of a tale that many medics know of but very few know properly. It was a genuine honour to have been able to talk with Dr James K. Styner about his incredible story and the birth of the Advanced Trauma Life Support courses. Dr Styner was generous to a fault with his time and pointed me at a newly published account of that famous day's events, authored by his son Randal Styner. That book, titled *The Light of the Moon* (2012), gives a much fuller account of the horror of the plane crash and the determination that led to the establishment of a new standard in trauma care. When Jim and I finally spoke I thanked him, belatedly, for getting me through the worst of that terrible day in Soho.

Advanced Trauma Life Support Manual, 6th Edition. American College of Surgeons. 1997.

Baker, M.S. 'Military medical advances resulting from the conflict in Korea, part I: Systems advances that enhanced patient survival.' *Military Medicine*. 2012 Apr; 177(4): 423–9.

Brøchner, A.C. & Toft, P. 'Pathophysiology of the systemic inflammatory response after major accidental trauma.' *Scandinavian Journal of Trauma, Resuscitation and Emergency Medicine*. 2009; 17: 43.

Buncombe, A., Marks, K., Finn, G., Watson-Smyth, K., Gregoriadis, L., Thornton, P. & Hann, M. 'Two dead, 81 injured as nail bomb blasts gay pub in Soho.' *Independent*. 1 May 1999. http://www.independent.co.uk/news/two–dead–81–injured–as–nail–bomb–blasts–gay–pub–in–soho–1096580.html

Cooper, G.J. & Taylor, D.E. 'Biophysics of impact injury to the chest and abdomen.' *Journal of the Royal Army Medical Corps*. 1989 Jun; 135(2): 58–67.

Elster, E. 'Trauma and the immune response: strategies for success.' *Trauma*. 2007 Jun; 62(6 Suppl): 54–5.

Frykberg, E. R. & Tepas, J. J. '3rd terrorist bombings. Lessons learned from Belfast to Beirut.' *Annals of Surgery*. 1988 Nov; 208(5): 569–76.

Holt, R. 'Soho nail bomber to serve at least 50 years.' *Daily Telegraph*. 2 March 2007. http://www.telegraph.co.uk/news/uknews/1544276/Soho–nail–bomber–to–serve–at–least–50–years.html

Hull, J. B. 'Traumatic amputation by explosive blast: Pattern of injury in survivors.' *British Journal of Surgery*. 1992 Dec; 79(12): 1303–6.

Katz, A. M. 'Ernest Henry Starling, his predecessors, and the "Law of the Heart".' *Circulation*. 2002; 106(23): 2986–92.

King, B. & Jatoi, I. 'The mobile army surgical hospital (MASH): A military and surgical legacy.' *Journal of the National Medical Association*. 2005 May; 97(5): 648–56.

Lee, C. C., Marill, K. A., Carter, W. A. & Crupi, R. S. 'A current concept of trauma-induced multiorgan failure.' *Annals of Emergency Medicine*. 2001 Aug; 38(2): 170–6.

Ng, R. L., James, S. E., Philp, B., Floyd, D., Ross, D. A., Butler, P. E., Brough, M. D. & McGrouther, D. A. 'The Soho nail bomb: The UCH experience. *Annals of the Royal College of Surgeons of England*. 2001 Sep; 83(5): 297–301.

Rignault, D. P. 'Recent progress in surgery for the victims of disaster, terror-ism, and war.' *World Journal of Surgery*. 1992 Sep/Oct; 16(5): 885–887.

Skandalakis, P. N., Lainas, P., Zoras, O., Skandalakis, J. E. & Mirilas, P. ' "To afford the wounded speedy assistance": Dominique Jean Larrey and Napoleon.' *World Journal of Surgery*. 2006 Aug; 30(8): 1392–9.

Styner, J. K. 'The birth of Advanced Trauma Life Support (ATLS).' *The Surgeon*. 2006; 4(3): 163–165.

Tsukamoto, T., Chanthaphavong, R. S. & Pape, H. C. 'Current theories on the pathophysiology of multiple organ failure after trauma.' *Injury*. 2010 Jan; 41(1): 21–6.

Vasagar, J. 'Soho bomb victims tell of devastation as pub torn apart.' *Guardian*. 8 June 2000. http://www.guardian.co.uk/uk/2000/jun/08/uksecurity.jeevanvasagar

INTENSIVE CARE

Intensive care can claim to have had many origins. *The History of British Intensive Care*, published as part of a Wellcome Trust 'Witnesses to Twentieth Century Medicine' project, details many contributing factors outside of the events of Copenhagen in 1953. However, Bjorn Ibsen's efforts during that polio epidemic still appear to have been key to the proliferation of larger, better-organised units dedicated to the care of critically ill patients.

The story of the epidemic that swept through Mauritius was unknown to me before writing this book. Interviewing my own father about life in the fishing village of Grand Gaube led to genuinely unexpected personal discoveries, about his early life and the devastation that polio brought to the family.

Here I must also thank Dr Nicholas Hirsch, a consultant anaesthetist at the National Hospital for Neurology and Neurosurgery who had a hand in training me while I was a junior doctor and whose enthusiasm for the history of anaesthesia and intensive-care medicine sparked my own.

I met Charles Gomersall while we were lecturing together on a disaster management course for the European Society for Intensive Care Medicine. The SARS epidemic became an event which most clinicians learnt about only in abstraction through research articles. The number of deaths worldwide was mercifully small – thanks largely to the efforts of Carlo Urbani and his colleagues – but that statistic belies the frankly heroic experience of a handful of intensive-care units and hospitals across the world that bore the brunt of the outbreak. I am grateful to Professor Gomersall for taking the time to speak with me about those events.

Andersen, E.W. & Ibsen, B. 'The anaesthetic management of patients with poliomyelitis and respiratory paralysis.' *British Medical Journal*. 1954; 1(4865): 786–8.

Anon. Extract from 'SARS in Hong Kong: From experience to action.' *Australian Health Review*., 2003; 26(3): 22–5.

Berthelsen, P.G. & Cronqvist, M. 'The first intensive care unit in the world: Copenhagen 1953.' *Acta Anaesthesiologica Scandinavica*. 2003; 47(10): 1190–5.

Chan-Yeung, M. & Yu, W.C. 'Outbreak of severe acute respiratory syndrome in Hong Kong Special Administrative Region: Case report.' *British Medical Journal*. 2003 Apr; 326(7394): 850–2.

Cyranoski, D. 2003. 'China joins investigation of mystery pneumonia.' *Nature*, 422(6931): 459.

Fleck, F. 'Carlo Urbani.' *British Medical Journal*. 2003 Apr; 326(7393): 825.

Fleck, F. 'How SARS changed the world in less than six months.' *Bulletin of the World Health Organization*. 2003. Available at: http://www.scielosp. org/scielo.php?script=sci_arttext&pid=s0042–96862003000800014.

McFarlan, A.M., Dick, G.W. & Seddon, H.J. 'The epidemiology of the 1945 outbreak of poliomyelitis in Mauritius.' *Quarterly Journal of Medicine*. 1946; 15: 183–208.

Peiris, J.S. et al. 'The severe acute respiratory syndrome.' *New England Journal of Medicine*. 2003; 349(25): 2431–41.

Peiris, J. S. M., Lai, S. T., Poon, L. L. M. et al. 'Coronavirus as a possible cause of severe acute respiratory syndrome.' *The Lancet*. 2003; 361: 1319–25.

Reilley, B., Van Herp, M., Sermand & Dentico, N. 'SARS and Carlo Urbani.' *New England Journal of Medicine*. 2003 May; 348(20): 1951–2.

Reisner-Sénélar, L., The Danish anaesthesiologist Björn Ibsen a pioneer of long-term ventilation on the upper airways. Dept. of Medicine. Johann Wolfgang Goethe University. 2009.

Reynolds, L.A. & Tansey, E.M. (eds). 'History of British intensive care c.1950–c.2000.' *Wellcome Witnesses to Twentieth Century Medicine*. Vol. 42. Queen Mary, University of London. 2011.

Richmond, C. 'Bjørn Ibsen.' *British Medical Journal*. 2007 Sept; 335(7621): 674.

Sample, D. & Evans, C. 'Estimates of the infection rates for poliomyelitis virus in the years preceding the poliomyelitis epidemics of 1916 in New York and 1945 on Mauritius.' *Journal of Hygiene*. 1957 June; 55(2): 254–65.

Wong, T. et al. 'Cluster of SARS among medical students exposed to single patient, Hong Kong.' *Emerging Infectious Diseases*. 2004 Feb; 10(2): 269–76.

WATER

Dr C.J. Brooks' long-running investigation into the factors that conspire to make helicopter crashes at sea so difficult to survive are touched upon only briefly at the start of this chapter. I had the pleasure of running into Dr Brooks at a conference about risk management in London last year. He tells me that when he travels on helicopters, he tapes a piece of string from the exit door, along the floor, to the seat in which he's sitting, to make sure he can find his way out of the vehicle in the event of an emergency!

My friend Dr Mike Tipton, a thermal physiologist at Southampton University, answered many queries I had here and elsewhere in the book about the human body's responses to the extremes of high and low temperatures. He was so helpful that I have almost forgiven him for making me endure the cold-shock response first hand, in a chilly pool of water, that doubles as his laboratory for physiological experimentation.

Brooks, C., 'The human factors relating to escape and survival from helicopters ditching in water.' 1989. Available at: http://oai.dtic.mil/oai/oai?verb =getRecord&metadataPrefix=html&identifier=ADA215755

Brooks, C., MacDonald, C. & Donati, L. 'Civilian helicopter accidents into water: Analysis of 46 cases, 1979–2006.' *Aviation, Space, and Environmental Medicine*. 2008 Oct; 79(10): 935–40.

Cheung, S. S., D'Eon, N. J. & Brooks C. J. 'Breath-holding ability of offshore workers inadequate to ensure escape from ditched helicopters.' *Aviation, Space, and Environmental Medicine*. 2001 Oct; 72(10): 912–8.

Craig, A. B. 'Causes of loss of consciousness during underwater swimming.' *Journal of Applied Physiology*. 1961; 16: 583–6.

Craig, A. B & Medd, W. L. 'Oxygen consumption and carbon dioxide production during breath-hold diving.' *Journal of Applied Physiology*. 1968; 24: 190–202.

Craig, A. B. & Ware, D. E. 'Effect of immersion in water on vital capacity and residual volume of the lungs.' *Journal of Applied Physiology*. 1967 Oct; 23(4): 423–5.

Craig, A. B. 'Depth limits of breath hold diving (an example of Fennology).' *Respiration Physiology*. 1968; 5: 14–22.

Craig, A. B. 'Heart rate responses to apneic underwater diving and to breath holding in man.' *Journal of Applied Physiology*. 1963; 18: 854–62.

Fahlman, A. 'The pressure to understand the mechanism of lung compression and its effect on lung function.' *Journal of Applied Physiology*. 2008; 104(4): 907–8.

Ferretti, G. & Costa, M. 'Diversity in and adaptation to breath-hold diving in humans.' *Comparative Biochemistry and Physiology Part A: Molecular & Integrative Physiology*. 2003; 136(1): 205–13.

Koehle, M., Lepawsky, M. & McKenzie, D. 'Pulmonary oedema of immersion.' *Sports Medicine*. 2005; 35(3): 183–90.

Levett, D. Z. & Millar, I. L. 2008. 'Bubble trouble: A review of diving physiology and disease.' *Postgraduate Medical Journal*. 2008; 84(997): 571–8.

Lindholm, P. & Lundgren, C. 'The physiology and pathophysiology of human breath-hold diving.' *Journal of Applied Physiology*. 2009 Jan; 106(1): 284–92.

Parkes, M. J. 'Breath-holding and its breakpoint.' *Experimental Physiology*. 2006 Jan; 91(1): 1–15.

Qvist, J. et al. 'Arterial blood gas tensions during breath-hold diving in the Korean ama.' *Journal of Applied Physiology*. 1993; 75(1): 285–93.

Rahn, H. 'Breath-hold diving: A brief history.' National Sea Grant Library. 2004. Available at: http://nsgl.gso.uri.edu/nysgi/nysgiw85001/nysgiw85001_part1.pdf

Schagatay, E. et al. 'Selected contribution: Role of spleen emptying in prolonging apneas in humans.' *Journal of Applied Physiology*. 2001; 90(4): 1623–9, discussion 1606.

Tipton, M. & Golden, F. 'Essentials of Sea Survival.' *Human Kinetics*. 2002.

ORBIT

Soyeon Yi literally rocketed to fame after her selection and flight as South Korea's first astronaut. I first noticed her at a space medicine conference in Houston. Welded to her smartphone, displaying boundless enthusiasm and constantly posting Twitter updates, she was never your average astronaut. She kindly agreed to let me interview her for this book and was good enough to check the story over afterwards to make sure of the detail.

I have watched three shuttle launches. All of them make you hold your breath. It was a privilege to be there for the launch and landing of the last shuttle mission – STS 135 – in July 2011. Among all of the potential insults that could be hurled at the human body, the energies involved in launch always made my focus, on the medical problems and the physiology, seem pretty irrelevant.

'Report of the Columbia Accident Investigation Board.' 2003. Washington. Available at: http://www.zsf.jcu.cz/jab/1_s1/supplementJAB1_1.pdf?set_language=cs

'Report of the Presidential Commission on the Space Shuttle Challenger Accident.' NASA. 6 June 1986.

Burroughs, W. *This New Ocean: The Story of the First Space Age.* Random House, 1999.

Curtis, H.D. *Orbital Mechanics for Engineering Students.* Elsevier, 2005; 257–73.

Houtchens, B.A. 'Medical-care systems for long-duration space missions.' *Clinical Chemistry.*1993; 39(1): 13–21.

Roth, E.M. 'Rapid (explosive) decompression emergencies in pressure-suited subjects.' NASA Contractor Report 1223. National Aeronautics and Space Administration. 1968; 1–125. Available at: http://eutils.ncbi. nlm.nih.gov/entrez/eutils/elink.fcgi?dbfrom=pubmed&id=5305515&ret mode=ref&cmd=prlinks

Summers, R.L. et al. 'Emergencies in space.' *Annals of Emergency Medicine.* 2005; 46(2): 177–84.

Wolfe, T. *The Right Stuff.* Jonathan Cape, 1979.

MARS

In 1997 we were definitely going to Mars. At least that's the way it looked to me. I remember that first visit to Houston very fondly. The hydrophonics and regenerative life support system experiments were run by a scientist by the name of Doug Ming, who reminded me a little of Bruce Dern's character in the film *Silent Running*.

Astrophysics was attractive to me because it was about boundary condition problems: observing the behaviour of systems at the extremes. Mars is in many ways the boundary condition mission for the human body.

I was the lucky recipient of a NESTA (National Endowment for Science Technology and the Arts) Fellowship and later a Bogue Fellowship that allowed me to continue my work with NASA. Professor Bill Paloski patiently facilitated my returns to Johnson Space Center and has served as both mentor and good friend across the years.

Human missions to Mars continue to sound like the stuff of

science fiction. But we will go to Mars before this century is out. Of that I am sure. If the end of the twenty-first century is as different from its beginning as the twentieth proved to be, we'll manage that and much, much more.

'Planetary exploration utilizing a manned flight system.' Office of Manned Space Flight, NASA Headquarters, Washington DC, 3 October 1966: 16.

SEI Synthesis Group. 'America at the threshold: America's space exploration initiative.' Washington DC: Government Printing Office, May 1991.

Carmeliet, G., Nys, G. & Bouillon, R. 'Microgravity reduces the differentiation of human osteoblastic MG–63 cells.' *Journal of Bone and Mineral Research*. 1997 May; 12(5): 786–94.

Carmeliet, G., Vico, L. & Bouillon, R. 'Space flight: A challenge for normal bone homeostasis.' *Critical Reviews. TM in Eukaryotic Gene Expression*. 2001; 11(1–3): 131–44.

Cassenti, B. 'Trajectory options for manned Mars missions.' *Journal of Spacecraft and Rockets*. 2005 Sept/Oct; 42(5).

Crocco, Gaetano, A. 'One-Year Exploration-Trip Earth–Mars–Venus–Earth.' Paper presented at the Seventh Congress of the International Astronautical Federation, Rome, Italy, 1956; 227–52.

D'Aunno, D.S., Dougherty AH, DeBlock HF & Meck JV. 'Effect of short- and long-duration spaceflight on QTc intervals in healthy astronauts.' *American Journal of Cardiology*. 2003 Feb; 91(4): 494–7.

Davis, J.R., Vanderploeg, J.M., Santy, P. A., Jennings, R.T. & Stewart, D. F. 'Space motion sickness during 24 flights of the space shuttle.' *Aviation, Space, and Environmental Medicine*. 1988 Dec; 59(12): 1185–9.

Drake, B. 'Reference Mission Version 3.0 Addendum to the Human Exploration of Mars: The Reference Mission of the NASA Mars Exploration Study Team.' 1998. Document ID: NASA/SP–6107–ADD.

Fitts, R.H., Riley, D.R. & Widrick, J.J. 'Functional and structural adaptations of skeletal muscle to microgravity.' *Journal of Experimental Biology*. 2001 Sep; 204(18): 3201–8.

Fogleman, G., Leveton, L. & Charles, J. 'The Bioastronautics Roadmap: A Risk Reduction Strategy for Human Exploration.' 1st Space Exploration Conference: Continuing the Voyage of Discovery, Orlando, Florida, 30 Jan 2005. AIAA–2005–2526.

Fritsch-Yelle, J.M., Leuenberger, U.A., D'Aunno, D.S., Rossum. A.C., Brown, T.E. & Wood, M.L. 'An episode of ventricular tachycardia during long–duration spaceflight.' *American Journal of Cardiology*. 1998 Jun; 81(11): 1391–2.

Graybiel, A., Clark, B., & Zarriello, J.J. 'Observations on human subjects living in a "slow rotation room" for periods of two days.' *Archives of Neurology*. 1960; 3: 55–73.

Holstein, G.R., Kukielka, E. & Martinelli, G.P. 'Anatomical observations of the rat cerebellar nodulus after 24 hr of spaceflight.' *Journal of Gravitational Physiology*. 1999 Jul; 6(1): 47–50.

Harm, D.L. & Parker, D.E. 'Preflight adaptation training for spatial orientation and space motion sickness.' *Journal of Clinical Pharmacology*. 1994 Jun; 34(6): 618–27.

Hohmann, W. *The Attainability of Heavenly Bodies* (translation of *Die Erreichbarkeit der Himmelskörper*). NASA Technical Translation F–44, Washington DC. 1960.

Joosten, K.B., 'Preliminary Assessment of Artificial Gravity Impacts to Deep-Space Vehicle Design.' NASA Johnson Space Center. 2007. NASA Document ID: 20070023306.

Lackner, J.R. 'Spatial orientation in weightless environments.' *Perception*. 1992; 21(6): 803–12.

LeBlanc, A., Schneider, V., Shackelford, L., West, S., Oganov, V., Bakulin,

A. & Voronin, A. 'Bone mineral and lean tissue loss after long duration space flight.' *Journal of Musculoskeletal and Neuronal Interactions*. 2000; 1(2): 157–60.

LeBlanc, A.D., Driscol, T.B., Shackelford, L.C., Evans, H.J., Rianon, N.J., Smith, S.M., Feeback, D.L. & Lai, D. 'Alendronate as an effective countermeasure to disuse induced bone loss.' *Journal of Musculoskeletal and Neuronal Interactions*. 2002; 2(4): 335–43.

Macho, L., Kvetnansky, R., Fickova, M., Popova, I.A. & Grigoriev, A. 'Effects of exposure to space flight on endocrine regulations in experimental animals.' *Endocrine Regulations*. 2001 Jun; 35(2): 101–14.

Oberth, H. 'Aufl.v. "Die Rakete zu den Planetenräume" kleines Stempel auf Vorsatz, sauberes Ex.' *Wege zur Raumschiffahrt* [*Ways to Spaceflight*]. München, 1929; xi, 423.

Paloski, W.H., Black, F.O. & Metter, E.J. 'Postflight balance control recovery in an elderly astronaut: A case report.' *Otology & Neurotology*. 2004 Jan; 25(1): 53–6.

Paloski, W.H. & Young, L.R. Artificial Gravity Worskhop. League City, Texas, USA in: *Proceedings and recommendations*, NASA Johnson Space Center and National Space Biomedical Research Institute, Houston, 1999.

Paloski, W.H. & Young, L.R. 'Artificial Gravity as a Multi-System Countermeasure to Bed Rest Deconditioning Pilot Study Overview.' 28th Annual International Gravitational Physiology. NASA Technical Reports Server. 8–13 Apr 2007; San Antonio, TX.

Paloski, W.H., Moore, S.T., Feiveson, A.H. & Taylor, L.C. 'Effects of Artificial Gravity and Bed Rest on Spatial Orientation and Balance Control.' NASA Johnson Space Center. NASA Technical Reports Server. 2007. Document ID: 20070011623.

Parker, D.E., Reschke, M.F., Arrott, A.P., Homick, J.L. & Lichtenberg, B. K. 'Otolith tilt-translation reinterpretation following prolonged weightlessness: Implications for preflight training.' *Aviation, Space, and Environmental Medicine*. 1985 Jun; 56(6): 601–6.

Portree, David, S. F., *Humans to Mars: Fifty Years of Mission Planning, 1950– 2000*. NASA Monographs in Aerospace History Series, Number 21, Feb 2001. (Excellent overview of American plans for sending men to Mars.)

Reschke, M. F., Kozlovskaya, I. B., Somers, J. T., Kornilova, L.N., Paloski, W. H. & Berthoz, A. J. 'Smooth pursuit deficits in space flights of variable length.' *Journal of Gravitational Physiology*. 2002 Jul; 9(1): 133–6.

Shi, S.J., South, D.A. & Meck, J.V. 'Fludrocortisone does not prevent orthostatic hypotension in astronauts after spaceflight.' *Aviation, Space, and Environmental Medicine*. 2004 Mar; 75(3): 235–9.

Solder, J. K, 'Round trip Mars trajectories: New variations on classic mission profiles.' American Institute of Aeronautics and Astronautics. 1990 Aug; Paper 90–3794.

Sonnenfeld, G. & Shearer, W.T. 'Immune function during space flight.' *Nutrition*. 2002 Oct; 18(10): 899–903.

Tsiolkovsky, K. E. 'Exploration of global space with jets' in *Collected Works*. Vol. 2: Nauka, Moscow, 1953: 100–139.

Turner, R. T. 'What do we know about the effects of space flight on bone?' *Journal of Applied Physiology*. 2000 Feb; 89: 870–47.

Vico, L, Collet, P., Guignandon, A., Lafage-Proust, M.H., Thomas, T., Rehaillia, M. & Alexandre, C. 'Effects of long-term microgravity exposure on cancellous and cortical weight-bearing bones of cosmonauts.' *The Lancet.* 2000 May; 355(9215): 1607–11.

Von, Braun, W. *The Mars Project*. University of Illinois Press, 1991.

Von, Braun, W. 'Crossing the Last Frontier,' *Collier's*. 22 March 1952: 24–31.

Walberg, G. 'How shall we go to Mars?' *Journal of Spacecraft and Rockets*. 1993 Mar/Apr; 30(2).

Waters, W.W., Ziegler, M.G. & Meck, J.V. 'Postspaceflight orthostatic hypotension occurs mostly in women and is predicted by low vascular resistance.' *Journal of Applied Physiology*. 2002 Feb; 92(2): 586–94.

Zubrin, R., Baker, D. & Gwynne, O. 'Mars Direct: A Simple Robust and Cost Effective Architecture for the Space Exploration Initiative.' American Institute of Aeronautics and Astronautics, 29th Aerospace Sciences Meeting, Reno, NV. 7–10 Jan 1991; 28. AIAA P.91–0328.

Zubrin, R. & Weaver, D.B. 'Practical Methods for Near-Term Piloted Mars Missions.' AIAA, SAE, ASME, and ASEE, Joint Propulsion Conference and Exhibit, Monterey, CA. 28–30 June, 1993. AIAA P.93–2089.

FINAL FRONTIERS

I will never forget James Hudson; he was so full of life even during what was to be his last admission. Over the weeks that I looked after him he told me the story of his remarkable life in instalments and always implored me to come back so that he could tell me more. I would like to thank his daughter Valerie Russell for giving me permission to recount some of the stories that I had been told as part of this book.

Mr Hudson had also contributed to a television documentary – *Lloyd George's War* – for the BBC in 1998, about his experiences during the First World War. The unedited footage from that programme is kept at the Imperial War Museum archives and it is from these that the story of his fumbled hand grenade comes. Additionally the British Dental Association holds a number of records and articles that allowed me to trace his career more accurately, from apprentice to qualified dentist and finally to hospital surgeon.

As junior I learnt something of the trade of elderly care medicine at the Hammersmith Hospital from my then registrar Dr Geoff Cloud, who is now a consultant in Stroke Medicine at St George's Hospital. When it came to writing this chapter, I visited him again to ask his advice. In that conversation and in further reading I got the impression that the field of gerontology is changing rapidly and, as elsewhere in medicine, our expectations have changed beyond any recognition. I hadn't previously appreciated the enormous complexity of the debate surrounding the search for a general theory of ageing. I was only able to scratch the surface of that here but include all of the material for further reading.

Caspari, R. & Lee, S. 'Older age becomes common late in human evolution.' *Proceedings of the National Academy of Sciences of the United States of America.* 2004; 101(30): 10895–900.

Chamberlain, G. 'British maternal mortality in the 19th and early 20th centuries.' *Journal of the Royal Society of Medicine.* 2006; 99(11): 559–63.

Christensen, K. et al. 'Ageing populations: The challenges ahead.' *The Lancet.* 2009; 374(9696): 1196–208.

Di, Biase, D. & Shelley, D., James Hudson, Obituary. *British Dental Journal.* 8 Sept 2001; 191(5): 282.

Gavrilov, L. A. & Gavrilova, N.S., 2003. 'The quest for a general theory of aging and longevity.' *Science of Aging Knowledge Environment.* 2003; (28): RE5.

Gems, D.H. & la Guardia, Y.D., 2012. 'Alternative perspectives on aging in *Caenorhabditis elegans*: Reactive oxygen species or hyperfunction?' *Antioxidants & Redox Signaling*. 2012 Sept; 1–29.

Griffin, J. 'Changing life expectancy throughout history.' *Journal of the Royal Society of Medicine*. 2008; 101(12): 577.

Hayflick, L. 'Entropy explains aging, genetic determinism explains longevity, and undefined terminology explains misunderstanding both.' *Public Library of Science Genetics*. 2007; 3(12): 220.

Hudson, J.A. 'The Dental Department at Redhill Hospital, Edgware as a criterion for the hospital services of the National Health Service.' *British Dental Journal*. 1948 Mar; 84: 100–2.

'Hudson receives fellowship at 102.' *British Dental Journal* (News and Notes), 2000; 189(10): 589.

Hudson, J.A. 'Long in the tooth: Witness for the profession.' *British Dental Journal*. 1991 Sept; 171(5): 138–40.

Kirkwood, T.B. 'Longevity in perspective.' *The Lancet*. 2006, 367(9511): 641–42.

Kirkwood, T.B. 1997. 'The origins of human ageing.' *Philosophical Transactions of the Royal Society of London*. Series B, Biological sciences. 1997; 352(1363): 1765–72.

Kirkwood, T.B. 'Where will it all end?' *The Lancet*. 2001; 357(9256): 576.

Kirkwood, T.B. *Time of Our Lives*. Phoenix, 2000.

Langdon, J. 'Mr James A. Hudson.' *Annals of the Royal College of Surgeons of England*. (Suppl) 2001; 83:26–27.

Lloyd George's War. BBC Timewatch. 1998. [Unedited interviews for programme, 1998, held by the Imperial War Museum; Catalogue Number TV 112X.]

Montagu, J.D. 'Length of life in the ancient world: A controlled study.' *Journal of the Royal Society of Medicine*. 1994; 87(1): 25–6.

Shay, J.W. & Wright, W.E. 'Hayflick, his limit, and cellular ageing.' *Nature* (reviews). 2000; 1(1): 72–6.

Walsh, J. 'The last soldier.' *Independent*. Review. 11 November 1999: 1.

ACKNOWLEDGEMENTS

THERE ARE MANY without whom this book would not have been possible. Foremost among them: Will Francis for chasing me for so many years to write something and Rupert Lancaster at Hodder for having faith in me and the idea. I had the great fortune to be in the hands of brilliant and eternally patient editor Tara Gladden. Her contributions vastly improved the text and I only hope she can forgive my many crimes against grammar and punctuation. Thanks also to the forever smiling Assistant Editor, Kate Miles who – somehow – always made me feel like things were running smoothly. For the inspired cover art my gratitude goes to Alice Laurent and Cliff Webb who did things with my heart and lungs that I never before thought possible. And thanks also to Kerry Hood on the publicity side who somehow managed the same with my diary.

I must finally say a huge thank you to the many people who have helped me along the way. It occurred to me that author and evolutionary biochemist Nick Lane, having bashed the idea of the importance of mitochondria and bioenergetics into my head over beers in the Jeremy Bentham pub, played a more important role than perhaps he knew. And I'm particularly grateful to those who read, commented and corrected, among them Mike Herd, Adam Rutherford, Mark Paul, Viki Mitchell and Neil Nixon. And none of this could have happened without Sue Rider and Sophie Kingston-Smith who are always there trying to manage the chaos that is the rest of my life.

Perhaps the biggest sacrifice was made by my wife Dee and our boys, who have given up weekends, evenings and holidays for too long to let me get to the end of this. Special thanks to the Wellcome Trust, the BBC and particularly to my colleagues and friends at University College London Hospital for allowing me to continue to explore and of course to my parents for encouraging me to do so in the first place.

PICTURE ACKNOWLEDGEMENTS

© AFP/Getty Images: 137. © Bettmann/Corbis: 109.
© Digitized Sky Survey 2/ Davide De Martin: 279.
© Kevin Fong: 247. Reproduced by kind permission
of the Guinea Pig Club: 35.© Michael Pitts: 175.
© Popperfoto/Getty Images: 9, 77.
Photograph by Art Rogers/Copyright © 1977,
Los Angeles Times, reprinted with permission: 211.

INDEX

Page numbers in *italic* refer to the illustrations

An invitation from the publisher

Join us at www.hodder.co.uk, or follow us
on Twitter @hodderbooks to be a part of
our community of people who love the very
best in books and reading.

Whether you want to discover more about a book
or an author, watch trailers and interviews, have the
chance to win early limited editions, or simply browse
our expert readers' selection of the very best books,
we think you'll find what you're looking for.

And if you don't, that's the place to tell us what's missing.

We love what we do, and we'd love you to be a part of it.

www.hodder.co.uk

 @hodderbooks

 HodderBooks

 HodderBooks